Sedgwick William Thompson

General Biology

Part 1: Introductory

Sedgwick William Thompson

General Biology
Part 1: Introductory

ISBN/EAN: 9783337216955

Printed in Europe, USA, Canada, Australia, Japan

Cover: Foto ©berggeist007 / pixelio.de

More available books at **www.hansebooks.com**

AMERICAN SCIENCE SERIES

GENERAL BIOLOGY

BY

WILLIAM T. SEDGWICK, Ph.D.

ASSOCIATE PROFESSOR OF BIOLOGY IN THE MASSACHUSETTS INSTITUTE OF TECHNOLOGY

AND

EDMUND B. WILSON, Ph.D.

PROFESSOR OF BIOLOGY IN BRYN MAWR COLLEGE

PART I.—INTRODUCTORY

NEW YORK

HENRY HOLT AND COMPANY

1886

Printed by
C. F. Vogelius.
New York.

PREFACE.

SEVERAL years ago it was our good fortune to follow as graduate students a course of lectures and practical study in General Biology, under the direction of Professor Martin, at Johns Hopkins University. Since that time we have ever been strongly of the opinion that beginners in Biology should be introduced to the subject by some similar method, following in the main the outlines marked out by Huxley and Martin more than ten years ago. The present work thus owes its origin to the influence of the authors of the "Elementary Biology," our deep indebtedness to whom we gratefully acknowledge.

It has not been our ambition to prepare an exhaustive treatise. We have sought only to lead beginners in biology from familiar facts to a better knowledge of how living things are built and how they act, such as may rightly take a place in general education or may afford a basis for further studies in General Biology, Zoölogy, Botany, Physiology, or Medicine.

It is still an open question whether the beginner should pursue the logical but difficult course of working upwards from the simple to the complex, or adopt the easier and more practical method of working downwards from familiar higher forms. Every teacher of the subject knows how great are the practical difficulties besetting the novice, who, provided for the first time with a compound microscope, is confronted with Yeast, Protococcus, or Amœba; and on the other hand, how hard it is to sift out what is general and essential from the heterogeneous details of a mammal or a flowering plant. In the hope of lessening the practical difficulties of the logical method we venture to submit a course of preliminary study, which we have used for some time with our own classes, and have found practical and effective. Believing that biology should follow the example of physics and chemistry in discussing at the outset the fundamental properties of matter and energy, we have devoted the first four chapters to an elementary account of living matter and vital energy. In the six chapters which follow, these facts are applied by a fairly exhaustive study of a representative plant and animal, of considerable, though not extreme, complexity—a method which we believe affords, in a given time, a better knowledge of vital phenomena than can be acquired by more superficial study of a larger number of forms. We are satisfied that the fern and earthworm are for this purpose the best available organisms, and that their study can be made fruitful and interesting. The last chapter comprises a brief account of the principles and outlines of classification as a guide in subsequent studies.

After this introductory study the student will be well prepared to take up the one-celled organisms, and can pass rapidly over the ground covered by such works as Huxley and Martin's *Practical Biology*, Brooks's *Handbook of Invertebrate Zoölogy*, Arthur, Barnes and Coulter's *Plant Dissection*, or the second part of this book, which is well in hand and will probably be ready in the course of the following year.

The directions for practical study are intended as suggestions, not substitutes, for individual effort. We have striven to make the work useful as well in the class-room as in the laboratory, and to this end have introduced many illustrations. The generosity of a friend has enabled us to enlist the skill of our friend Mr. James H. Emerton, who has drawn most of the original figures from nature, under our direction. We have also been greatly aided in the preparation of the figures by Mr. William Claus of Boston.

NEW YORK, September, 1886.

TABLE OF CONTENTS.

CHAPTER X.

THE BIOLOGY OF AN ANIMAL.—*(continued.)*

CHAPTER XI.

OUTLINES OF CLASSIFICATION.

APPENDIX.

CHAPTER I.

INTRODUCTORY.

We know from common experience that all material things are either dead or alive, or, more accurately, that all matter is either lifeless or living: and so far as we know, life exists only as a manifestation of living matter. Living matter and lifeless matter are everywhere totally distinct, though often closely associated. The most careful studies have on the whole rendered the distinction more clear and striking, and have demonstrated that living matter never arises spontaneously from lifeless matter, but only through the immediate influence of living matter already existing. And so, whatever may have been the case at an earlier period of the earth's history, we are at present justified in regarding the line between living and lifeless matter as one of the most clearly defined and remarkable of natural boundaries.

The contrast between living matter and lifeless matter is made the ground for a division of the natural sciences into two corresponding groups, viz.: the **Biological Sciences** and the **Abiological Sciences**, dealing with living matter and lifeless matter respectively. The biological sciences (p. 7) are known collectively as **Biology** (*βίος, life : λόγος, a discourse*), which is therefore often defined as the science of life, of living things, or of living matter. But living matter, as will shortly appear, is only ordinary matter which has entered into a peculiar state or condition. And hence biology is more precisely defined as *the science which treats of matter in the living state.*

The Relationship between Living and Lifeless Matter. Although living matter and lifeless matter present this remarkable contrast to one another, they are most intimately related, as a moment's reflection will show. The living substance of the human body, or of any animal or plant, is only the transformed lifeless matter of the food which has been taken into the body and has there assumed, for a time, the living state.   Lifeless matter in the shape of food is continually streaming into all living things on the one hand and passing out again as waste on the other.   In its journey through the organism some of this matter enters into the living state and lingers for a time as part of the bodily substance.   But sooner or later it dies, and is then for the most part cast out of the body (though a part may be retained within it, either as an accumulation of waste material, or to serve some useful purpose).   Matter may thus pass from the lifeless into the living state and back again to the lifeless, over and over in never-ending cycles.   A living plant or animal is like a whirlpool into which, and out of which, matter is constantly streaming, while the whirlpool maintains its characteristic form and individuality.*

How then is living matter different from lifeless matter?   The question cannot be fully answered by chemical analysis, for the reason that this process necessarily kills living matter, and the

---

* We append Huxley's graphic comparison of a living organism to a whirlpool (*The Crayfish as an Introduction to the Study of Zoölogy*, pp. 84, 85.   Appleton, N. Y., 1880), having made slight verbal changes so that the passage, originally applied to a particular animal, the crayfish, may be made general:—

"To put the matter in the most general shape, the body of the organism is a sort of focus to which certain material particles converge, in which they move for a time, and from which they are afterward expelled in new combinations.   The parallel between a whirlpool in a stream and a living being, which has often been drawn, is as just as it is striking.   The whirlpool is permanent, but the particles of water which constitute it are incessantly changing.   Those which enter it on the one side are whirled around and temporarily constitute a part of its individuality; and as they leave it on the other side, their places are made good by newcomers.

"Those who have seen the wonderful whirlpool, three miles below the Falls of Niagara, will not have forgotten the heaped-up wave which tumbles and tosses, a very embodiment of restless energy, where the swift stream hurrying from the Falls is compelled to make a sudden turn toward Lake Ontario.   However changeful in the contour of its crest, this wave has been visible, approximately in the same place and with the same general form, for centuries past.   Seen from a mile off it would appear to be a stationary hillock of water.   Viewed closely it is a typical expression of the conflicting impulses generated by a swift rush of material particles.

"Now, with all our appliances, we cannot get within a good many miles, so to speak, of the living organism.   If we could, we should see that it was nothing but the constant form of a similar turmoil of material molecules, which are constantly flowing into the organism on the one side and streaming out on the other."

results therefore teach us little of the chemical conditions existing in the matter when alive. Analyses, nevertheless, bring to light several highly important facts. It appears that living matter is a tolerably definite compound of a number of the chemical elements, and it is probably too low an estimate to say that at least six elements must unite in order that life may exist. Moreover, only a very few out of all the elements are able, under any circumstances, to form this living partnership.

The most significant fact, however, is that there is no loss of weight when living matter is killed. The total weight of the lifeless products is exactly equal to the weight of the living substance analyzed, and if anything has escaped at death it is imponderable, and, having no weight, is not material. It follows that living matter contains no material substance peculiar to itself, and that every element found in living matter may be found also, under other circumstances, in lifeless matter.

Considerations like these lead us to recognize a fundamental fact, namely, that the terms living and lifeless designate two different STATES or CONDITIONS of matter. We do not know, at present, what causes this difference of condition. But so far as the evidence shows, the living state is never assumed except under the influence of antecedent living matter, which, so to speak, infects lifeless matter, and in some way causes it to assume the living state.

Distinctive Properties of Living Matter. Those properties of living matter which, taken together, distinguish it absolutely from every form of lifeless matter, are:

1. Its chemical composition.
2. Its power of waste and repair, and of growth.
3. Its power of reproduction.

Living matter invariably contains substances known as **proteids**, which are believed to constitute its essential material basis (see p. 35). Proteids are complex compounds of Carbon, Oxygen, Hydrogen, Nitrogen, Sulphur, and (in some cases at any rate) Phosphorus.

It has been frequently pointed out that each of these six elements is remarkable in some way: oxygen, for its vigorous combining powers; nitrogen, for its chemical inertia; hydrogen, for its great molecular mobility; carbon, sulphur, and phosphorus, for their allotropic properties, etc. All of these peculiarities may be shown to be of significance when considered as attributes of living matter. (See Herbert Spencer, *Principles of Biology*, vol. i.)

It is not, however, the mere presence of proteids which is characteristic of living matter. White-of-egg (albumen) contains an abundance of a typical proteid and yet is absolutely lifeless. Living matter does not simply contain proteids, but has the *power to manufacture them* out of other substances; and this is a property of living matter exclusively.

The powers of living matter are still more characteristic. It is continually wasting away by a kind of internal combustion, but continually repairs the waste by the processes of growth. Moreover, this growth is of a characteristic kind, differing absolutely from the so-called growth of lifeless things. Crystals and other lifeless bodies grow, if at all, by *accretion*, or the addition of new particles to the outside. Living matter grows from within by *intus-susception*, or taking in new particles, and fitting them into the interstices between those already present, throughout the whole mass. And, lastly, living matter not only thus repairs its own waste, but also gives rise by reproduction to new masses of living matter which become detached from the parent mass and enter forthwith upon an independent existence.

We may perceive how extraordinary these properties are by supposing a locomotive engine to possess like powers: to carry on a process of self-repair in order to compensate for wear, to grow and increase in size, detaching from itself at intervals pieces of brass or iron endowed with the power of growing up step by step into other locomotives capable of running themselves, and of reproducing new locomotives in their turn. Precisely these things are done by every living thing, and nothing in any degree comparable with them takes place in the lifeless world.

Huxley has given the best statement extant of the distinctive properties of living matter, as follows:—

"1. Its *chemical composition*—containing, as it invariably does, one or more forms of a complex compound of carbon, hydrogen, oxygen, and nitrogen, the so-called protein (which has never yet been obtained except as a product of living bodies), united with a large proportion of water, and forming the chief constituent of a substance which, in its primary unmodified state, is known as protoplasm.

"2. Its *universal disintegration and waste by oxidation*, and its concomitant *reintegration by the intussusception of new matter*. A process of waste resulting from the decomposition of the molecules of the protoplasm in virtue of which they break up into more highly oxidated products, which cease to form any part of the living body, is a constant concomitant of life. There is reason to believe that carbonic acid is always one of these waste products, while the others contain the remainder of the carbon, the nitrogen, the hydrogen, and the other elements which may enter into the composition of the protoplasm.

"The new matter taken in to make good this constant loss is either a ready-formed protoplasmic material, supplied by some other living being, or it consists of the elements of protoplasm, united together in simpler combinations, which constantly have to be built up into protoplasm by the agency of the living matter itself. In either case, the addition of molecules to those which already existed takes place, not at the surface of the living mass, but by interposition between the existing molecules of the latter. If the processes of disintegration and of reconstruction which characterize life balance one another, the size of the mass of living matter remains stationary, while if the reconstructive process is the more rapid, the living body *grows*. But the increase of size which constitutes growth is the result of a process of molecular intus-susception, and therefore differs altogether from the process of growth by accretion, which may be observed in crystals, and is effected purely by the external addition of new matter; so that, in the well-known aphorism of Linnæus, the word 'grow' as applied to stones signifies a totally different process from what is called 'growth' in plants and animals.

"3. Its *tendency to undergo cyclical changes*. In the ordinary course of nature, all living matter proceeds from pre-existing living matter, a portion of the latter being detached and acquiring an independent existence. The new form takes on the characters of that from which it arose; exhibits the same power of propagating itself by means of an offshoot; and, sooner or later, like its predecessor, ceases to live, and is resolved into more highly oxidated compounds of its elements.

"Thus an individual living body is not only constantly changing its substance, but its size and form are undergoing continual modifications, the end of which is the death and decay of that individual; the continuation of the kind being secured by the detachment of portions which tend to run through the same cycle of forms as the parent. No forms of matter which are either not living or have not been derived from living matter exhibit these three properties, nor any approach to the remarkable phenomena defined under the second and third heads." (*Encyclopædia Britannica*, 9th ed., art. "Biology," vol. iii., p. 679.)

Concerning the chemical composition of living matter and its significance, a few considerations should be borne in mind. For the purposes of biological study life must be regarded as a property of a certain kind of compounded matter.* But we are

---

* Herbert Spencer, *Principles of Biology*, vol. 1.

forced to regard the properties of compounds as the resultants of the properties of their constituent elements, even though we cannot well imagine how such a connection exists; and so in the long run we have to fall back upon the properties of carbon, hydrogen, nitrogen, oxygen, etc., for the properties of living matter. Reflections of this sort show how ignorant we are of the real properties of the elements, and how important further study of them is.

Consciousness. It cannot be doubted that the living form of matter alone manifests the phenomena of consciousness; and it is not generally believed to exist to any great extent, except in the higher forms of animal life. Because of the purely subjective nature of consciousness it is impossible to say with absolute certainty whether animals other than ourselves are really conscious, and many of the actions commonly supposed to indicate consciousness are simply reflex actions, even in ourselves.[*] We do not know, therefore, how widely consciousness is distributed among living things, though there is no reason to suppose that plants possess it. There is evidence, however, that it exists in different degrees of intensity among animals, keeping pace in a general way with the amount of organization which they exhibit, and shading off into simpler nervous conditions. The phenomena of consciousness are undoubtedly the least understood of all biological problems. From the scientific point of view it appears at present impossible to translate them into the terms of physical phenomena, though they are not known to occur apart from a living material basis with which they appear to be in some way closely connected.

Scope of Biology. Biology has already been defined as the science which treats of matter in the living state. Whatever pertains to living matter or to living things pertains to biology. The study of the forms of living things, of their habits, actions, nutrition, surroundings, distribution; of their relations to the lifeless world, and to one another; of their sensations and social relations, their origin and their fate, these and many other topics

---

[*] " It is really an open question whether a crayfish has a mind or not. Moreover the problem is an absolutely insoluble one inasmuch as nothing short of being a crayfish would give us positive assurance that such an animal possesses consciousness." (Huxley, *The Crayfish*, p. 89.)

lie within the province of biology. The science includes both
zoölogy and botany, dealing with the phenomena of animal
and vegetal life not only separately, but in their relations to one
another. The term **General Biology** does not designate a particular member of the group of biological sciences, but is only
a convenient phrase, which has recently come into use for the
general introductory study of biology. It includes a description of the general properties of living matter as revealed in the
structures and actions of living things, and may serve as the
basis for subsequent study of more special branches of the
science. It deals with the broad characteristic phenomena and
laws of life as they are illustrated by the thorough comparative study of a series of plants and animals taken as representative types; but in this study the student should never
lose sight of the fact that all the varied phenomena which may
come under his observation are in the last analysis due to *the
properties of matter in the living state*, and that this matter and
these properties are the real goal of the study.

The Biological Sciences. The field covered by biology as thus
understood is so wide as to necessitate a subdivision of the subject
into a number of branches which are usually assigned the rank of
distinct sciences. They are arranged in a tabular view on p. 8.
The table shows two different ways of dividing the main subject,
according as the table is read from left to right or *vice versa*. According to the more usual arrangement biology is primarily divided
into zoölogy and botany, according as animals or plants respectively, form the subject of study. Such a division has the great
advantage of practical convenience since, as a matter of fact, most
biologists devote their attention mainly either to plants alone, or
to animals alone. From a scientific point of view, however, a better
subdivision is into *Morphology* (μορφή, *form*; λόγος, *a discourse*)
and *Physiology* (φύσις, *nature*; λόγος, *a discourse*). The former
is based upon the facts of form, structure, and arrangement, and
is essentially statical; the latter upon those of action or function,
and is essentially dynamical. But morphology and physiology are
so intimately related that it is impossible to separate either subject
absolutely from the other.

Besides the sub-sciences given upon the table a distinct branch
called *Ætiology* is often recognized, having for its object the in-
vestigation of the causes of biological phenomena. But the

Biology.

The science of living things; i.e., of matter in the living state.

Morphology.

The science of form, structure, etc. Essentially statical.

Anatomy.

The science of structure; the term being usually applied to the coarser and more obvious composition of plants or animals.

Histology.

Microscopical anatomy. The ultimate optical analysis of structure by the aid of the microscope; separated from anatomy only as a matter of convenience.

Taxonomy.

The classification of living things. Based chiefly on phenomena of structure.

Distribution.

Considers the position of living things in space and time, their distribution over the present face of the earth and their distribution and succession at former periods, as displayed in fossil remains.

Embryology.

The science of development from the germ. Includes many mixed problems pertaining both to morphology and physiology. At present largely morphological.

Physiology.

The science of action or function. Essentially dynamical.

Physiology.

The special science of the functions of the individual in health and in disease; hence including *Pathology.*

Psychology.

The science of mental phenomena.

Sociology.

The science of social life, i.e., the life of communities, whether of men or of lower animals.

Botany.

The science of vegetal living matter or plants.

Biology.

The science of living things; i.e., of matter in the living state.

Zoology.

The science of animal living matter or animals.

scientific study of every phenomenon has for its ultimate object the discovery of its cause. Ætiology is therefore inseparable from any of the several branches of biology and cannot properly be assigned an independent place.

Psychology and *Sociology* are not yet generally admitted to constitute branches of biology, and it is customary and convenient to set them apart from it. The progress of these sciences has clearly shown, however, that their study is inseparable from that of biology in the ordinary sense.

It will be observed in the scheme that morphology and physiology overlap; that is, there are certain biological sciences in which the study of structure and of action cannot be separated. This is especially true of embryology, which considers the successive stages of embryonic structure and also the modes of action by which they are produced. And finally it must not be forgotten that any particular arrangement of the biological sciences must be in the main a matter of convenience only; for it is impossible to study any one order of phenomena in complete isolation from all others.

PRACTICAL STUDY.*

A. THE PARTS OF THE MICROSCOPE.

    a. The Stand.
        1. Stage.
        2. Mirror.
        3. Diaphragms.
        4. Coarse and fine Adjustments.

    b. The Optical Apparatus.
        1. Objective.
        2. Ocular or Eye-piece.
        3. Draw-tube.

B. THE USE OF THE MICROSCOPE.

        1. Illuminating.
        2. Focussing.
        3. Camera lucida.
        4. Measurements by the camera and the stage micrometer.
        5. The value of binocular vision.

C. THE EXAMINATION AND SKETCHING OF VARIOUS OBJECTS AS PRACTICE.

D. SIMPLE METHODS OF TEASING, MOUNTING, AND STAINING.

E. THE PRINCIPLES OF PHOTO-MICROGRAPHY.

---

* Since a knowledge of the microscope is an indispensable preliminary to the laboratory work, the student may now conveniently learn something of its use.

CHAPTER II.

THE COMPOSITION OF LIVING ORGANISMS.

Living and Lifeless Matter in the Living Organism. Since our own bodies and those of lower animals and of plants are composed of matter, it might be supposed, from what has been said in the last chapter, that they are composed of living matter. This, however, is true only in part. It is strictly true that every plant or animal contains living matter, but a little reflection will show that it contains lifeless matter also. In the human body lifeless matter is found in the hairs, the ends of the nails, and the outer layers of the skin,—structures which are not simply devoid of feeling, as every one knows them to be, but are really lifeless in every sense, although forming part of a living body. Nor is lifeless matter confined to the exterior of the body. The mineral matter of the bones is not alive; and this is true, though less obviously, of many other parts, such as the liquid basis of the blood, or plasma, the fat (which is never wholly absent) and various other forms of matter occurring in many parts of the body.

In lower animals examples of this truth occur on every hand. The calcareous shells of animals like the snail and the oyster; the skeletons of corals and sponges; the hard outer crust of insects, lobsters, and related animals; the scales of fish and reptiles; the feathers, claws, and beaks of birds; the fur of animals; these are a few of the countless instances of structures composed wholly or in part of lifeless matter, which nevertheless enter into the composition of living animals.

Among plants like facts are even more conspicuous. No one can doubt that the outer bark of an oak is devoid of life. The heart-wood of a tree is entirely dead, and even in the so-called live wood, through which the sap flows, not only is the solid part of the wood lifeless, but also the sap itself.

In short, it is a general truth that the living matter of plants and animals is almost invariably associated with lifeless matter;

and therefore the first step to be taken in the study of biology is
to examine how the matter of a living organism is arranged.
This may be done conveniently by a preliminary study of the

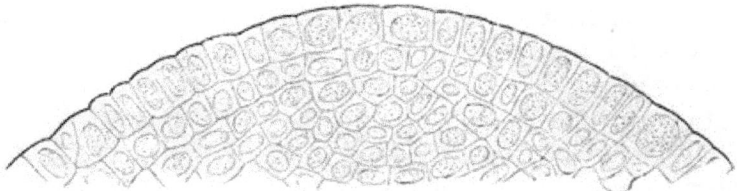

FIG. 1. (After Sachs.)—Longitudinal section through the growing apex of a young pine-
shoot. The dotted portion represents the protoplasm, the narrow lines being the partition walls composed of cellulose $(C_6H_{10}O_5)$. (Highly magnified.)

general composition of living organisms, leaving out of consideration, for the present, the lowest and simplest forms, and con-
sidering only those more highly organized plants and animals
which are in a greater or less degree familiar to every one.

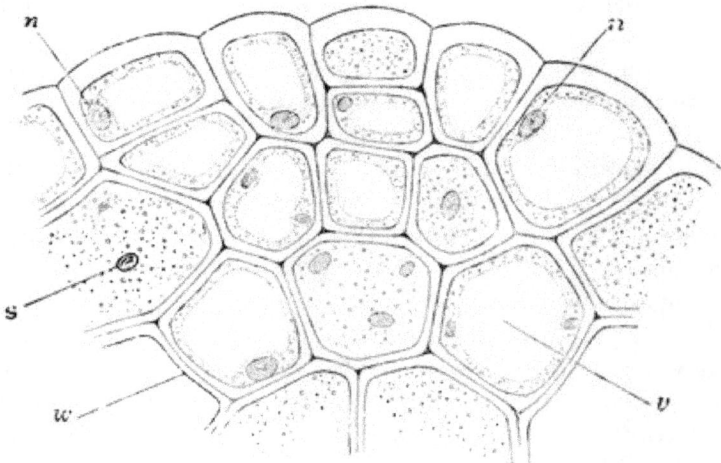

FIG. 2.—Cross-section through part of the young leaf of a fern (*Pteris aquilina*), showing
thick-walled cells; most of the walls are double. The granular substance is protoplasm.
Most of the cells contain a large central cavity (vacuole) filled with sap, the protoplasm
having been reduced to a thin layer inside the partitions. Nuclei are shown in some of
the cells, and lifeless grains of starch in others: *n*, nuclei; *s*, starch; *v*, vacuole; *w*, double
partition-wall. (× 300.)

Organisms composed of Organs. Functions. It is characteristic
of any living body—for example, a rabbit or a geranium—that it
is composed of unlike parts, whose structure enables them to

perform various operations essential or accessory to the life of the whole. The plant has stem, roots, branches, leaves, stamens, pistil, seeds, etc.; the animal has externally head, trunk, limbs, eyes, ears, etc., and internally stomach, intestines, liver, lungs, heart, brain, and many other parts having the most diverse structures. These parts are known as *organs*, and the living body is called an *organism* because it possesses them.

The word organism, as here used, applies only to the higher animals and plants. It will be seen in the sequel that there are forms of life so simple that organs as here defined can scarcely be distinguished or are entirely wanting. Such living things are nevertheless called organisms because of the existence of different parts which are analogous in function to the organs of plants and animals.

Since organisms are composed of unlike parts, they are said to be heterogeneous in structure. They are also heterogeneous in action, the different organs performing different operations called *functions*. For instance, it is the function of the stomach to digest food, of the heart to pump the blood into the vessels, of the kidneys to excrete waste matters from the blood, and of the brain to direct the functions of other organs. A similar diversity of functions exists in plants. The roots hold the plant fast and absorb various substances from the soil; the stem supports the leaves and flowers and conducts the sap; the leaves breathe and elaborate portions of the food;

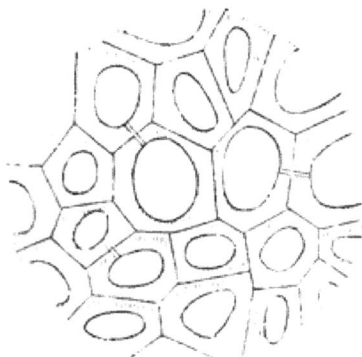

FIG. 3. (After Sachs.) — Cross-section through a group of dead, thick-walled wood-cells from the stem of maize. The cells contain only air or water. (Highly magnified.)

and the reproductive organs of the flower serve to form and bring to maturity seeds destined to give rise to a new generation.

Heterogeneity of the kind just indicated is one of the most characteristic features of living things, and is not known in any mass of lifeless matter, however large and complex.

Organs composed of Tissues. Differentiation. In the next place, it is to be observed that the organs themselves, when fully formed, are not homogeneous, but are in turn made up of different parts. The human hand is an organ which consists of many parts, differ

ing widely in structure and function.   Outside are the skin, the
hairs, the nails; inside are bones, muscles, tendons, ligaments,
blood-vessels, and nerves.   The leaf of a plant is an organ consisting of a woody framework (the veins) which supports a green pulp,
the whole being covered on the outside by a delicate transparent
skin.   In like manner every organ of a higher plant or animal
may be resolved into different parts, and these are known as

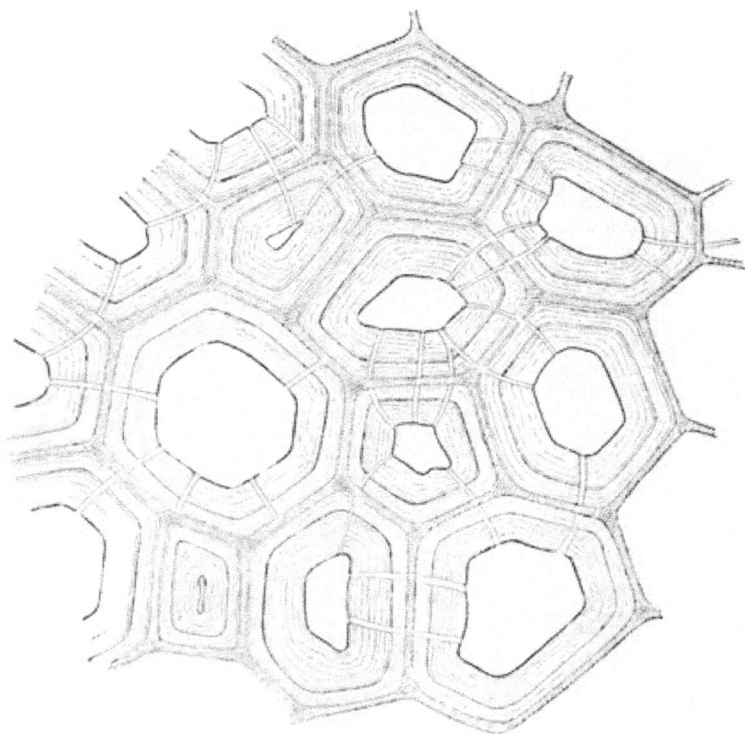

FIG. 1.—Cross-section through dead wood-like cells from the underground stem of a fern
(*Pteris aquilina*).   The walls are uncommonly thick and the protoplasm has disappeared.
(× 50.)

tissues.   The tissues of fully formed organs are often very different from one another, as in the cases just mentioned; that is, they
are well *differentiated;* but frequently in adult organs, and
always in those which are sufficiently young, the tissues shade
gradually into one another, so that no definite line can be drawn
between them.   In such cases they are said to be less differentiated.   For example, in the full-grown leaf of a plant the woody

framework, the green cells, and the skin exist as three plainly different tissues. But in younger leaves these same tissues are less different, and in very young leaves, still in the bud, there are no visible differences and the whole organ is very nearly homogeneous. In this case the tissues are *undifferentiated*, though potentially capable of differentiation. In the same way, the tissues of the embryonic human hand are imperfectly differentiated, and at a very early stage are undifferentiated.

Tissues composed of Cells. Finally, microscopical examination shows every tissue to be composed of minute parts known as *cells*, which are nearly or quite similar to one another throughout the whole tissue, and form the ultimate units out of which the tissues and organs, and hence the whole organism, are built, somewhat as a house is built of bricks or pieces of wood.

It will be shown in Chapter IV, that the ultimate units or cells possess everywhere the same fundamental structure; but they differ immensely in form, size, and mode of action, not only in different animals and plants, but even in different parts of the same individual. As a rule, the cells of any given tissue are closely similar one to another and are devoted to the same function, but differ from those of other tissues in form, size, arrangement, and especially in function. Indeed, the differences between tissues are merely the outcome of the differences between the cells composing them. The skin of the hand differs in appearance and uses from the muscle which it covers, because skin-cells differ from muscle-cells in form, size, color, function, etc. Hence a tissue may be defined as *a group of similar cells having a similar function.** As a rule, each organ consists of several such groups of cells or tissues, but, as stated above, young organs are nearly or quite homogeneous; that is, all of the cells are nearly or quite alike. It is only when the organ grows older that the cells become different and arrange themselves in different groups,—a process known as the *differentiation of the tissues.* In the case of some organs—for instance the leaf of a moss—the cells always remain nearly alike, somewhat as in the embryonic condition, and the whole organ consists of a single tissue.

---

* Tissues frequently contain matters deposited between cells, but these have been directly derived from the cells, and vary as the cells vary.

What has been said thus far applies only to higher plants and animals. But it is an interesting and suggestive fact that there are also innumerable isolated cells, both vegetal and animal, which are able to carry on an independent existence as one-celled plants or animals. Physiologically these must certainly be regarded as individuals; but it is no less certain that they are equivalent, morphologically, to the constituent cells of ordinary many-celled organisms. It will appear hereafter that the study of such unicellular organisms forms the logical groundwork of all biological science.

Since organisms may be resolved successively into organs, tissues, and cells, it is evident that cells must contain living matter. And a cell may be defined as a *small mass of living matter either living apart or forming one of the ultimate units of an organism.*

Lifeless Matter in the Living Tissues. In the tissues the living cells are not, as a rule, in immediate contact one with another, but are more or less completely separated by partitions of lifeless matter. This may be seen in a section through some rapidly growing organ like a young shoot (Fig. 1). The whole mass is formed of nearly similar closely crowded cells separated by very narrow partitions. Each cell consists of a mass of granular, viscid, living substance known as *protoplasm*, part of which is condensed into a more solid, large, and rounded body, the *nucleus.*

In such a group of cells no tissues can be distinguished; or, rather, the whole mass consists of a single tissue (meristem), which is almost entirely composed of living matter (protoplasm). In older tissues the partitions often increase in thickness, as shown in Fig. 2. In every case *the*

B

A

FIG. 5. (After Ranvier.)—Muscle-cells; *A*, from the intestine of a dog, in cross-section; *B*, single isolated cell, from the intestine of a rabbit, viewed from the side. (× 350.)

partitions are composed of lifeless matter which has been manufactured and deposited by the living protoplasm constituting the bodies of the cells. In still older parts of the plant certain of the lifeless walls may become extremely thick, the protoplasm entirely disappears, and the whole tissue (wood) consists of life-

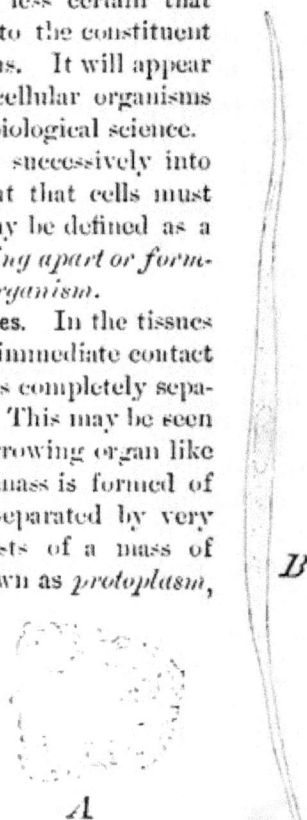

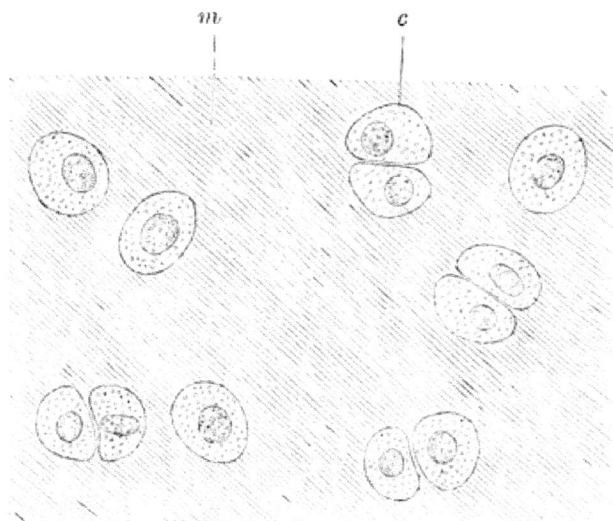

Fig 6. (After Schäfer)—Human cartilage (from head of metatarsal bone); *c*, cells; *m*, lifeless matrix. (× 600.)

Fig. 7. (Modified from Ranvier)—Blood of frog, showing two forms of cells (corpuscles), one flattened and oval, one branched. (× 650.)

less matter enclosing spaces filled with air or water (Figs. 3 and 4).

Among animals analogous cases are common. The muscles of the small intestine, for instance, (Fig. 5), consist of bundles of elongated cells (*fibres*) each of which is composed of living matter surrounded by a very thin covering (*sheath*) of lifeless matter. In cartilage or gristle, which covers the ends of many bones (Fig. 6), the oval cells are very widely separated by the deposition between them of large quantities of solid lifeless matter forming what is known as the *matrix*. In blood (Fig. 7) the flattened or irregular cells (*corpuscles*) are separated by a lifeless fluid (*plasma*) in which they float. In bone (Fig. 8) the cells have a branching, irregular form, and are separated by solid calcareous matter which is unmistakably lifeless. These examples show that the lifeless matters of the body often occur in the form of deposits between living cells by which they have been produced. In all such cases the embryonic tissue consists at first of living cells in direct contact, or separated by only a very small quantity of lifeless matter. In later stages the cells may manufacture additional lifeless

FIG. 8. (Modified from Schenk.)—Section of bone from the human femur showing the living branching bone-cells lying in the bony lifeless matrix. Diagrammatic.

substance which appears in the form of solid partition-walls between the cells, or as a matrix, solid or liquid, in which the cells lie.

Equally important is the formation of lifeless matter *within* cells, either (*a*) by the deposition of various substances in the protoplasm, or (*b*) by the direct transformation of the whole mass of protoplasm. Examples of the first kind are mineral crystals, (Fig. 9), grains of starch (Fig. 9), drops of water, and many other substances found within the cells of plants. Among animals drops of fat (Fig. 10) and calcareous or siliceous deposits are similarly produced. Indeed, there is scarcely any limit to the number of lifeless substances which may thus appear within the cells both of plants and animals.

The second case is of less importance, though of common oc-

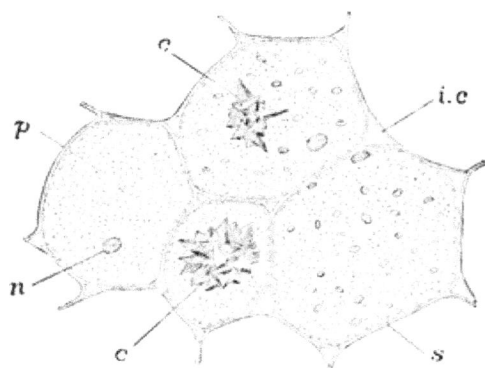

Fig. 9.—A group of cells from the stem of a geranium (*Pelargonium*), showing lifeless substances (starch and crystals) within the protoplasm. As in Fig. 2, each cell contains a large central vacuole, filled with sap; *c*, groups of crystals of calcium oxalate; *i.c*, intercellular space; *n*, nucleus; *s*, granules of starch. (× 300.)

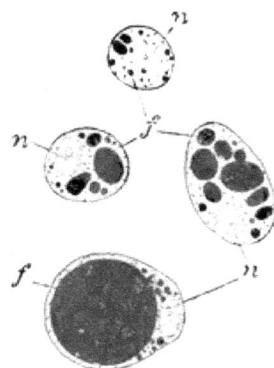

Fig. 10. (After Ranvier.)—Group of "adipose cells" from the tissue beneath the skin ("subcutaneous connective tissue") of an embryo calf, showing drops of fat in the protoplasm; *f*, fat drops (black); *n*, nuclei. (× 550.)

currence. A good example is found in the lining membrane of the œsophagus of the dog (Fig. 11), which like the human skin is almost entirely made up of closely crowded cells. Those in the

Fig. 11.—Section through the inner coat of the gullet of a dog, showing: *p*, living cells of the deeper layers; *s*, lifeless cells of the superficial layers.

deepest part consist chiefly of living protoplasm very similar to that of the young pine shoot (compare Fig. 1). Above them the cells gradually become flattened until at the surface they have the form of flat scales. As the cells become flattened their substance changes. The protoplasm diminishes in quantity and dies; so that near the surface the cells are wholly dead, and finally fall off.* In a similar manner are formed the lifeless parts of nails, claws, beaks, feathers, and many related structures. A hair is composed of cells essentially like those of the skin. At the root of the hair they are alive, but as they are pushed outwards by continued growth at the root, they are transformed bodily into a dead, horny substance forming the free portion of the hair. Feathers are only a complicated kind of hairs and are formed in the same way.

It is a significant fact that the quantity of lifeless matter in the organism tends to increase with age. The very young plant or animal probably possesses a maximum proportion of protoplasm, and as life progresses lifeless matter gradually accumulates within or about it,—sometimes for support, as in tree-trunks and bony skeletons; sometimes for protection as in oyster- and snail-shells; sometimes apparently from sheer inability on the part of the protoplasm to get rid of it. Thus we see that youth is the period of life and vigor, and age the period of comparative lifelessness, in a literal and material as well as in a figurative sense.

Summary. The bodies of higher animals and plants are heterogeneous in structure and function, and are composed of organs. These may be resolved into one or more tissues, each of which consists of a mass of similar cells (or their derivatives) having a similar function. The cells are small masses of living matter, or protoplasm, which deposit more or less lifeless matter either around (outside) them or within their substance. In the former case the protoplasm may continue to live, or it may die and be absorbed. In the latter case it may likewise live on for a time, or may die, either disappearing altogether or leaving behind a residue of lifeless matter.

---

* It is probable that this process is connected with the deposition within the protoplasm of extremely minute granules of lifeless matter (*keidin*).

PRACTICAL STUDY.

The student should understand, once for all, that the principal points observed must be recorded by notes and by *sketches*, good or bad, whether he can "draw" or not. Sketches should be first made in outline, should be simple and *on a large scale*. They can afterwards be made very clear by slightly shading the different parts with colored pencils. The aim should be to represent the *natural relations of parts* rather than their minute details, and accidental displacements should be disregarded.

A. Organs.

1. Examine and enumerate the more important organs of some familiar dissected animal, e.g., a cat, frog, or pigeon. Make a sketch showing the position of these organs, appending a list of their functions so far as you can determine them from your general knowledge.

2. Make a like examination of some familiar flowering plant; e.g., a lily or geranium. What general resemblances, and what differences, between plant and animal do you find?

B. Tissues.

1. Prepare (or examine) the leg, or some other organ, of a mammal or frog so dissected as to show some of the component tissues. Learn their names from the demonstrator and make a sketch showing the appearance and relative position of at least three of them.

2. Study the leaf of a plant. Observe the following tissues:
  a. Skin or *epidermis*, which, after moistening, may easily be scraped off with a knife.
  b. Supporting tissue—the network of *veins* or *nerves*.
  c. Mesophyll—the green pulp.

C. Cells.

1. Make with a sharp razor a thin slice or "section" of the stalk of a large leaf (or of a succulent young stem, e.g., of celery). Mount in water and study it with the microscope, at first with a low power, afterwards with a higher. Sketch some of the cells.

2. With a pair of forceps tear off a bit of epidermis from the upper surface of a leaf; mount in water, examine it with the microscope and sketch a few cells.

3. Examine with the microscope a drop of newt's or frog's blood. Sketch some of the cells (*corpuscles*).

4. Gently scrape the inside of your cheek with a scalpel. Mount the *débris* in water or 0.5% aqueous solution of sodium chloride (*normal fluid*), examine and sketch the cells.

5. Tease out a bit of frog's muscle in dilute glycerine. Examine and sketch the elongated cells (*fibres*).

D. Living and Lifeless Matter.

1. Mount in normal fluid a thin bit or section of cartilage (if possible from the frog's sternum). Observe the cells embedded in the matrix. Place a few drops of dilute iodine solution near one edge of the cover-glass and draw it under by applying a bit of blotting-paper to the opposite edge. Remove the excess of iodine by drawing through a drop of distilled water. The cells are killed and stained brown; the matrix hardly stains.

2. Prepare and mount a thin slice of a potato-tuber from *just below* the skin. Stain with very dilute iodine. The living protoplasm is killed and stained brown. The numerous lifeless starch-grains turn blue, or blue-black if the iodine is strong.

3. Prepare a slide of newt's or frog's blood as in C, 3. Stain with iodine and note the effect.

4. Make a thin transverse section of a geranium-stem (*Pelargonium*) and mount in water. Search for crystals of oxalate of calcium. When found, observe and sketch. Observe also the thin, lifeless partition-walls composed of cellulose.

5. Examine a prepared section of bone. The clear basis is composed of lifeless calcium-salts. The minute branching cavities (*lacuna*) were in life occupied by branching living cells (*bone-cells*).

CHAPTER III.

LIVING MATTER OR PROTOPLASM.

It is evident from the last chapter that protoplasm must be the starting-point for detailed studies of organisms such as we are soon to enter upon, or indeed for biological studies of any kind. The present chapter will therefore be devoted to a general account of protoplasm, beginning with a sketch of its discovery and afterward discussing its characteristic appearances and structure, its properties, chemical basis, and relations to various physical agents, such as heat, light, electricity, etc.

Historical Sketch. Protoplasm was observed long before its significance was understood. The discovery of its essential identity in plants and animals, and ultimately of the extreme importance of the rôle which it everywhere plays, must be reckoned as one of the greatest scientific achievements of this century.

Dujardin, in 1835, applied the name *sarcode* (σαρκώδη, *fleshy*) to certain forms of living matter forming the bodies of some of the simplest animal organisms; but the word *protoplasm* (πρῶτος, *first*; πλάσμα, *form*) was first used by Von Mohl in 1846 to designate a portion of the contents of vegetal cells, though he was ignorant of its full significance. In 1850 Cohn definitely maintained the essential identity of the "sarcode" of the zoölogists and the "protoplasm" of the botanists, a view which had already been propounded in different words by Payen four years before. Scientific research advanced rapidly to the discovery which now lies at the root of biology, viz.: that life is a manifestation of this remarkable substance, essentially identical in chemical, physical, and other properties, in all plants and animals. To Max Schultze belongs the credit of having given the first comprehensive statement of this identity (1860), and he is now generally recognized as the founder of biology in the modern sense. He first extended the meaning of the word protoplasm to include all living matter, both vegetal and animal, in which sense the word is now universally employed.

Structure. Protoplasm differs slightly in appearance in different plants and animals, and in different parts, and different stages

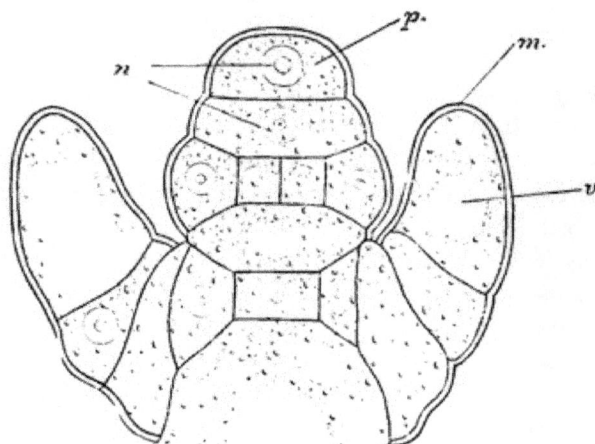

FIG. 12. (After Sachs.)—Young growing cells from the extreme tip of a stonewort (*Chara*):
m, membrane; *n*, nuclei; *p*, protoplasm; *v*, vacuole filled with sap.   (× 550.

of development, of the same individual.   The protoplasm of very
young structures is considered to be the simplest or most typical.

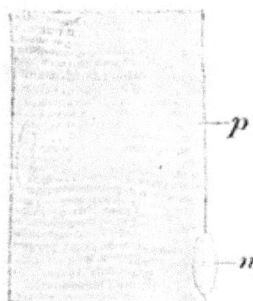

FIG. 13. (After Ranvier.)—
Part of a single fibre of voluntary muscle from the leg
of a rabbit. *p*, protoplasm;
n, nucleus. × 500.

—as it appears, for instance, in the eggs
of many animals, or in the cells of very
young vegetal shoots   Such protoplasm
is called *typical* or *unmodified*. As it
grows older it usually changes more or
less, both in chemical and physical properties, and in appearance.   In the cells of
a young organ (Figs. 1 and 12), protoplasm is a transparent, semi-fluid or viscid,
granular substance which has a watery
appearance but shows no tendency to
mix with water.   Treated with solutions of certain coloring matters (for
instance with iodine) it generally absorbs them from the solution and be-

comes deeply colored, often assuming a much darker shade
than that of the solution.   Under moderately high powers of

the microscope (500–1000 diameters) it appears to consist mainly of two constituents, viz.: (*a*) a perfectly clear semi-fluid basis, (*b*) minute, solid, highly refracting granules embedded in the clear substance. With the most perfect and powerful lenses skilled observers detect also fine threads or fibres in the clear substance, but these elude all ordinary observation. In most cases part of the protoplasm of the cell appears as a definite, rounded mass known as the *nucleus*, which is easily distinguishable from the remaining protoplasm by its darker appearance (due to its greater refractive power) and the deeper color which it assumes when treated with solutions of coloring mat-

ter. In spite of this apparently simple structure, we must infer from its mar-vellous powers that protoplasm is really endowed with a molecular constitution of extreme complexity, probably far exceeding that of any lifeless sub-stance.

Such protoplasm, simple and un-modified, is seldom found absolutely pure even in very young cells. It con-tains almost always an admixture of other substances which become more abundant as the protoplasm grows older, taking the shape of fluid, solid or even gaseous deposits. Common examples of such deposits are drops of water, oil, and resin, granules of pigment,

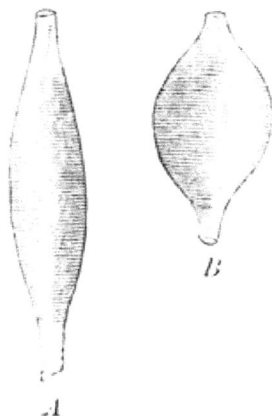

FIG. 14.—Change of form in a contracting muscle. *A*, muscle in the ordinary or extended state; *B*, the same muscle when contracted. (Diagram.)

starch and solid proteid matters, and crystals of mineral substances like calcium oxalate, phosphate and carbonate, and silica. Bub-bles of gas are said to appear rarely in the protoplasm; but this, if true, is certainly exceptional. The living substance itself often changes in appearance as the cells become differentiated. The protoplasm of voluntary muscles (Fig. 13) is firm, clear, non-granu-lar, highly refractive, and arranged in alternating bands or stripes of darker and lighter substance. In some cases (e.g. the outer portions of the skin, or of a hair, as explained in Chap. II.) the modifications of the cell-substance becomes so great that both its physical and chemical constitution are entirely altered, and it is no longer protoplasm, but some form of lifeless matter.

Protoplasm in Action. We know that living things are the seat of active changes, which taken together constitute their life. In the last analysis these changes are undoubtedly chemical actions taking place in the protoplasm, which in many cases do not produce visible results, but only molecular transformations. There is no doubt that extensive and probably very complex molecular actions go on in the protoplasm of young growing cells, though it may appear absolutely quiescent to the eye, even when this is armed with a powerful microscope. In other cases, the chemical action produces perceptible changes in the protoplasm,—for instance, some form of motion,—just as the invisible chemical action in an electrical battery may be made to produce visible effects (light, locomotion, etc.) through the agency of an electrical motor.

A familiar instance of protoplasmic movement is the contraction of a muscle. This process is most likely a change of molecular arrangement, causing the muscle, while keeping its exact bulk, to change its form, the two ends being brought nearer together (Fig. 14). The visible change of form is here supposed to be due to an invisible change of molecular arrangement, and this in turn to be produced by chemical action taking place in the living substance.

A striking and beautiful example of movement in protoplasm occurs in the simple organism known as *Amœba* (Fig. 15). The entire body of this animal consists of a mass of naked protoplasm enclosing a nucleus, or sometimes two; in other words, it is a single naked cell. The protoplasm of an active *Amœba* is in a state of ceaseless movement, contracting, expanding, flowing, and changing the form of the animal to such an extent that it is known as the " Proteus animalcule." The whole movement is a kind of flux. A portion of the protoplasm flows out from the mass, making one or more prolongations (*pseudopods*) into which the remainder of the protoplasm finally passes, so that the whole body advances in the direction of the flow. If particles of food be met with, the protoplasm flows around them, and when they have been digested within the body, the protoplasm flows onward, leaving the refuse behind. Hour after hour and day after day this flowing may go on, and there is perhaps no more fascinating and suggestive spectacle known to the biologist. A similar change of form is exhibited by the colorless corpuscles of amphibian and other blood, in which it may be observed, though far less satisfactorily, if *Amœba* cannot be obtained.

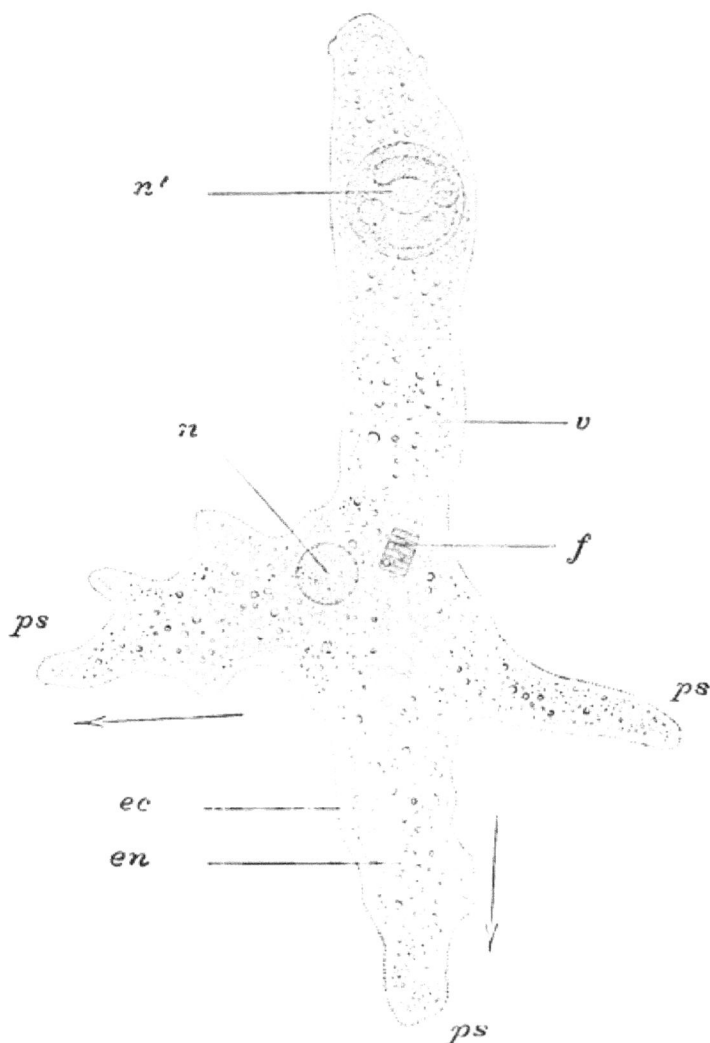

Fig. 15.—The Proteus animalcule (*Amœba Proteus*). The arrows indicate the direction of the flow. *f,* a microscopic plant which has been eaten; *n, n',* nuclei; *ps,* pseudopods; *v,* vacuole. (× 300).

Among plants, protoplasmic movements of perhaps equal beauty may be observed. One of the simplest is known as the *rotation* of protoplasm, which may be studied to advantage in rather young cells of stoneworts (*Chara* or *Nitella*). These cells have the form of short or elongated cylinders which are often pointed at one end (Fig. 16*a*). The protoplasm is surrounded by a delicate membrane which thus forms a sac enclosing the protoplasm. In very young cells the protoplasm entirely fills the sac; but as the cell grows older a drop of liquid appears near the centre of the mass and increases in size until the protoplasm is reduced to a thin layer (*primordial utricle*), lining the inner surface of the membrane (compare Fig. 2). In favorable cases the entire mass of protoplasm is seen to be flowing steadily around

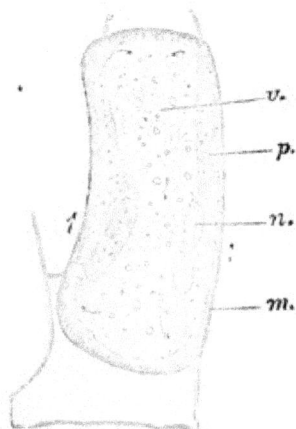

Fig. 16.—A cell of a stonewort (*Nitella*) showing the rotation of protoplasm; the arrows show the direction of the flow. *m*, membrane of the cell; *n*, nucleus, opposite to which is a second; *p*, protoplasm; *v*, large central vacuole filled with sap.

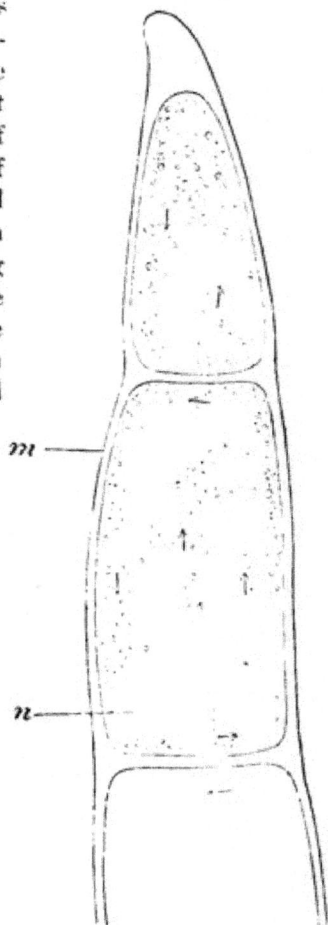

Fig. 16*a*.—Two cells and a part of a third from the tip of a " leaf " of a stonewort, showing rotation of the protoplasm in the direction of the arrows.

the inside of the sac as indicated by the arrows in Fig. 16. It moves upwards on one side, downwards on the opposite side, and in opposite directions across the ends, forming an unbroken circuit. The flow is rendered more conspicuous by various granules and other lifeless masses floating in the protoplasm and by the large oval nucleus or nuclei, all of which are swept onward by the current in its ceaseless round. A similar rotation of protoplasm occurs in many other vegetal cells, some of the best examples being certain leaf-cells of *Vallisneria* and *Anacharis*.

A second and somewhat more intricate kind of movement in vegetal protoplasm is known as *circulation*. This differs from rotation chiefly in the fact that the protoplasm travels not only in a peripheral stream but also in strands which run across through the central space (vacuole) and thus form a loose network. Circulation is well seen in cells composing the hairs of various

FIG. 17.—Flower-cluster (*a*) and single stamen (*b*) of a cultivated spiderwort (*Tradescantia*); *h*, hairs upon the stamen. *a*, slightly reduced; *b*, slightly enlarged.

plants, such as the common nettle (*Urtica*), the spiderwort (*Tradescantia*), the hollyhock (*Althœa*), and certain species of gourds (*Cucurbita*). It may be conveniently studied in the hairs upon the stamens of the cultivated spiderwort (*Tradescantia*). The flower of this plant is shown in Fig. 17*a* and one of the stamens with its hairs at *b*. Each hair consists of a single row of elongated cells covered by delicate membranes and connected by their ends. As in *Nitella*, the protoplasm does not fill the cavity of the sac, but forms a thin lining (*primordial utricle*) on its inner face (Fig. 18). From this layer delicate threads of protoplasm reach into and pass through the central cavity, where they often branch and are connected together so as to form a very loose network. The nucleus (*n*) is embedded

either in the peripheral layer or at some point in the network, and
the threads of the latter always converge more or less regularly to
it. In active cells currents continually flow to and fro throughout
the whole mass of protoplasm. In the threads of the network

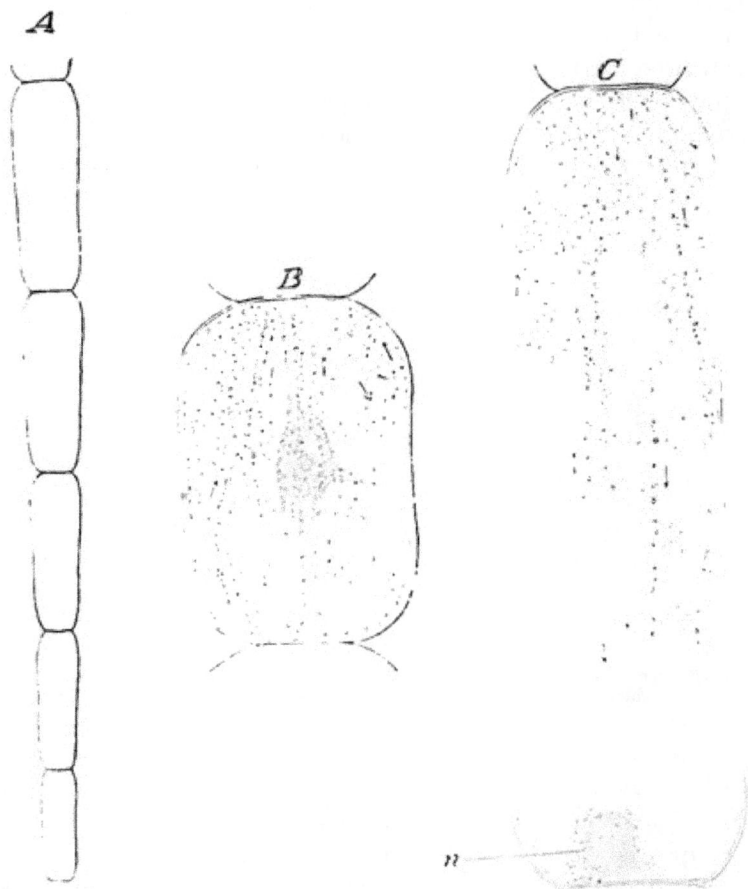

FIG. 18.—Enlarged cells of the hairs from the stamens of the spiderwort. *A*, five cells,
somewhat enlarged, protoplasm not shown; *B* and *C*, cells much more enlarged, showing
the circulation of protoplasm as indicated by the arrows; *n*, nucleus.

granules are borne rapidly along, gliding now in one direction,
now in another; and although the flow is usually in one direction
in any particular thread, no system can be discovered in the complicated movements of the whole. In the larger threads the
curious spectacle often appears of two rapid currents flowing in

opposite directions on opposite sides of the same thread. The currents in the thread may be seen to join currents of the peripheral layer which flow here and there, but without the regularity observed in the protoplasm of *Nitella*. The protoplasmic network, also, as a whole undergoes a slow but steady change of form, its delicate strands slowly swaying hither and thither, while the nucleus travels slowly from point to point.*

Finally, we may consider an example of a form of protoplasmic movement known as *ciliary* action, which plays an important rôle in our own lives and those of lower animals and of some plants. The interior of the trachea, or windpipe, is lined by cells having the form shown in Fig. 19. At the free surface of the cell (turned towards the cavity of the trachea) the protoplasm is produced into delicate vibratory filaments having a sickle-shape when bent; these are known as

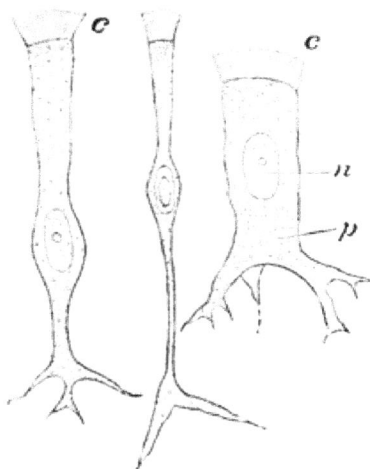

Fig. 19. (After Klein.—Three isolated ciliated cells from the interior of the windpipe of the cat; *c*, the cilia at the free end; *n*, the nucleus; *p*, the protoplasm. (Highly magnified.)

cilia (*cilium*, an eyelash). They are so small and lash so vigor-

---

* The similar circulation which may be seen in a nettle hair has been described by Huxley (*Lay Sermons*, etc., p. 123, American ed.): "The common nettle owes its stinging property to the innumerable stiff and needle-like, though exquisitely delicate, hairs which cover its surface. Each stinging needle tapers from a broad base to a slender summit, which though rounded at the end is of such microscopic fineness that it readily penetrates and breaks off in the skin. The whole hair consists of a very delicate outer case of wood, closely applied to the inner surface of which is a layer of semi-fluid matter, full of innumerable granules of extreme minuteness. This semi-fluid lining is protoplasm, which thus constitutes a kind of bag, full of a limpid liquid, and roughly corresponding in form with the interior of the hair which it fills. When viewed with a sufficiently high magnifying power, the protoplasmic layer of the nettle hair is seen to be in a condition of unceasing activity. Local contractions of the whole thickness of its substance pass slowly and gradually from point to point, and give rise to the appearance of progressive waves, just as the bending of successive stalks of corn by a breeze produces the apparent billows of a corn-field."

ously as to be nearly or quite invisible until the movements are in some way made sluggish. The movement is then seen to be more rapid and vigorous in one direction than in the other, all the cilia working together like the oars of a row-boat. By this action a definite current is produced in the surrounding medium (in this case the mucus of the trachea) flowing in the direction of the more vigorous movement. In the trachea this movement is upwards towards the mouth, and mucus, dust, etc., are thus removed from the lungs and windpipe. In many lower animals and plants, especially in the embryonic state, cilia are used as organs of locomotion, serving as oars to drive the organism through the water. The male reproductive germs of plants and animals are also propelled in a similar fashion.

Actions like those just described are known as *automatic*. They take place without evident cause and do not answer to any discoverable external stimulus. Automaticity in this sense is a property of all protoplasm; for it appears under numberless forms of which movement is only one example. Another general property of protoplasm which has been well illustrated by the phenomena just described is known as *contractility*. The swaying of the strands of protoplasm of *Tradescantia*, the shortening of the muscle-fibre, the waves in the protoplasmic current of *Nitella*, the protrusion of the pseudopods in *Amœba*, and the bending of the cilia in the windpipe are all due to this intrinsic property of living matter.

These examples of visible vital actions have been chosen to help the student to a better conception of the invisible phenomena which underlie them. There can be no doubt that within the substance of a living gland-cell or brain-cell there is a play of invisible molecular actions far more tumultuous and complicated than the visible movements displayed in *Nitella* or a nettlehair; and it is of the utmost importance that the student should attain to a full and vivid sense of the reality and energy of this invisible activity even in protoplasm which under the closest scrutiny appears to be absolutely quiescent.

Energy. Whence comes the power expended in protoplasmic action? The power to do work, or *energy*, is required for every action; and according to the doctrine of the conservation of energy it is never generated *de novo*, but only transformed or set free. Physicists distinguish two forms of energy, viz., *kinetic* energy, the energy of motion, and *potential* or *latent* energy, the energy of position. The former is active, the latter passive. What is called the liberation of energy consists simply in the transformation of potential into kinetic energy. Thus, a stone lifted above the ground possesses potential energy because its position enables it to fall, and in falling it can do work or overcome resistance. When the stone is let fall, its former energy of position becomes energy of motion; and in like manner every form of kinetic energy is produced by the liberation of potential energy.

Now the energy which protoplasm expends in doing work is **potential energy derived from its food.** Protoplasm takes in food-matters rich in energy and transforms them into its own substance, thus acquiring a store of intrinsic potential energy. This energy is ultimately set free by complex *chemical changes* taking place in the living protoplasm itself. Although these changes are not yet fully understood, they are known to consist, broadly speaking, in the breaking-down or decomposition of complex unstable compounds into simpler and stabler ones under the influence of oxygen; and in the long run they are complicated processes of combustion or oxidation (cf. p. 4). It is a well-known fact that energy is set free whenever strong chemical affinities are satisfied at the expense of weaker ones; that is, whenever the elements unite to form compounds, or whenever an unstable compound is resolved into one or more stabler ones.* In lifeless matter such processes may or may not be processes of oxidation; in living matter they are probably in the long run always processes of oxidation.

This will be rendered clearer by a few illustrations. The energy (or power) which drives a steam-engine is first set free in the furnace by the union of the fuel with the oxygen of the air—that is, by a process of oxidation in which the mutual affinities of oxygen and the carbon of the fuel are satisfied. The energy which propels a cannon-ball is potential in the gunpowder and becomes active at the moment of explosion. Gunpowder con-

---

* Conversely, kinetic energy is used up or made potential whenever stable compounds are converted into those less stable, as happens, for example, in the decomposition of carbon dioxide by sunlight in green plants.

tains two free elements, sulphur and carbon, and one compound, saltpetre or nitrate of potassium (KNO_3). Saltpetre is an unstable compound, the nitrogen, oxygen and potassium being loosely combined. When heat is applied in the form of a spark the oxygen is released from its feeble union and suddenly combines with the carbon and sulphur to form stabler compounds—namely, carbon dioxide (CO_2), monoxide (CO), etc.—the nitrogen being more or less completely set free. In other words, the strong affinities of oxygen for carbon and sulphur are satisfied at the expense of its weaker affinities for potassium and nitrogen, the result being a sudden liberation of energy which produces the explosion. Again, nitro-glycerine is a very complex compound, $C_3H_5O_3(NO_2)_3$, in a state of unstable equilibrium; that is, its elements are bound together by weak affinities which may readily be overcome. A sudden shock, such as a blow, causes an instantaneous and violent re-arrangement of the elements of the molecule into a number of simpler and stabler molecules (CO_2, CO, H_2O, N, etc.). Here, again, strong affinities, held in check under ordinary conditions, are suddenly satisfied at the expense of weaker affinities, and an enormous amount of energy is set free.[*]

Something of this sort we must picture to ourselves as taking place in living matter. Protoplasm takes in various complex food-matters and absorbs free oxygen. These substances are built up into the living fabric where they are loosely united to form unstable or explosive compounds containing a large amount of potential energy. By the timely and orderly explosion of these compounds, under nicely regulated conditions, their potential energy is converted into kinetic energy which manifests itself as the so-called *vital force* underlying every vital act. Apart from the phenomena of consciousness, vital actions thus conceived lose much of their supposed mysterious and unfathomable character; for "vital force" is not a unique and isolated power, but a form of physical energy derived from and convertible into other well-known forces of nature.

The foregoing considerations lead directly to an account of the chemical composition of living matter.

_____

[*] According to Berthelot, the reaction is as follows : $2C_3H_5O_3(NO_2)_3 = 6CO_2 + 5H_2O + 6N + O$.

Chemical Relations.

We know nothing of the precise chemical composition of living protoplasm, because, as has been said (p. 2), living protoplasm cannot be subjected to chemical analysis without destroying its life. But the results of chemical examinations leave no doubt that the molecules of protoplasm are highly complex and are probably separated from one another by layers of water.

Proteids. It has already been stated (p. 3) that the characteristic products of the analysis of protoplasm are the group of closely related substances known as *proteids*. But proteids form only a small part of the total weight of any plant or animal, being always associated with quantities of other substances. Even the white of an egg, which is usually taken for a typical proteid, contains only twelve per cent of actual proteid matter, the remainder consisting chiefly of water. The following table shows the percentage of proteids and other matters in a few familiar organisms and their products :

PROXIMATE PERCENTAGE COMPOSITION OF SOME COMMON SUBSTANCES.*

Arranged according to richness in Proteids.

		Water.	Proteids.	Carbo-hydrates.	Fats.	Other substances.
1	Apples.............................	84.8	0.4	14.3	0.0	0.5
2	Indian Corn, aerial portion fresh...	84.3	0.3	14.7	0.5	0.4
3	Oysters, shells included	15.4	1.0	0.6	0.2	82.8
4	Turnips	91.2	1.0	6.9	0.2	0.7
5	Melons	95.2	1.1	2.5	0.6	0.6
6	Sweet Potatoes	75.8	1.5	21.2	0.4	1.2
7	Crayfish, whole	10.0	1.9	0.1	0.1	87.9
8	Irish Potatoes	75.5	2.0	21.3	0.2	1.0
9	Clams, round, shells included......	27.3	2.1	1.3	0.1	69.2
10	Oats, aerial part fresh............	81.0	2.3	15.3	0.5	0.9
11	Grass " "	75.0	3.0	19.9	0.8	0.3
12	Peas " "	81.5	3.2	12.8	0.6	0.8
13	Cow's Milk	87.4	3.4	4.8	3.7	0.7
14	Flounder, whole...................	25.2	5.2	0.0	0.3	67.3
15	Lobster, "	22.0	5.4	0.2	0.5	60.9
16	Poplar and Elm leaves, fresh	50.0	6.0	22.0	1.5	4.5
17	Crab, whole	20.1	7.3	0.5	0.9	56.2
18	Brook Trout, whole................	69.2	9.0		1.1	18.7
19	Hen's Eggs, shells included	65.6	11.4	0.5	10.8	12.0
20	Mutton "chops"....................	41.5	12.5		29.3	16.9
21	Chicken, whole	42.2	14.3		1.1	12.4
22	Beef, heart.......................	73.4	14.9		24.8	6.9
23	Beef, liver	69.5	20.4	4.5	5.1	1.5
24	Beefsteak, round, lean............	69.0	20.5		8.1	12.2
25	Beans	12.7	23.4	57.1	2.1	3.6
26	Cheese	31.2	27.4	2.4	25.1	3.9
27	Cheese from skimmed milk	44.3	38.3	9.0	6.8	4.6

* Compiled chiefly from tables of food composition prepared by W. O. Atwater for the Smithsonian Institution, though a few examples have been added —viz., numbers 2, 10, 11, 12, 16,—from *Ingram's Hot Cross Grove*, N. Y., 1883.

All proteids have nearly the same chemical composition and similar physical properties, however different may be the forms of protoplasm in which they occur. The analysis of protoplasm, or rather of the proteids which are its basis, teaches us really nothing of its vital properties, but serves only to show the chemical composition of the material basis by which these are manifested.

Proteids are so called from their resemblance to protein ($\pi\rho\omega$-$\tau o s$, *first*), a substance first described and named by Mulder. According to Hoppe-Seyler they have approximately the following percentage composition:

	C.	H.	N.	O.	S.
From	51.5	6.9	15.2	20.9	6.3
to	54.5	7.3	17.0	23.5	2.0

A small quantity of phosphorus is also very frequently present. Associated with these elements are always small quantities of various mineral substances which remain as the ash when protoplasm is burned; but the nature of their relations to the other elements is uncertain. The ash varies both in quantity and chemical composition in different animals and plants. In the white-of-egg the chief constituents of the ash are potassium chloride (KCl) and sodium chloride (NaCl, the former being much in excess. The remainder consists of phosphates, sulphates, and carbonates of sodium and potassium, with minute quantities of calcium, magnesium, and iron, and a trace of silicon. Many other mineral substances occur in association with other kinds of proteids, but always in very small proportion. These salts are in some way essential to the activity of protoplasm, as we know by familiar experience. Man, like other animals and the plants, requires certain mineral substances (e.g. common salt), but we have no knowledge of the part these play in protoplasm.

It is important to note the substantial chemical identity of animal and vegetal proteids, because this is one reason for regarding vegetal and animal protoplasm as essentially identical in other respects. The following table from Johnson, after Gorup-Besanez and Ritthausen, shows the percentage composition of various proteids, and proves that the difference between vegetal and animal proteids is chemically no greater than that between different kinds of vegetal or different kinds of animal proteids :

PERCENTAGE COMPOSITION OF PROTEIDS.

	C.	H.	N.	O.	S.
Animal Albumen.............	53.5	7.0	15.5	22.4	1.6
Vegetal　　"　　...............	53.4	7.1	15.6	23.0	0.9
Animal Casein.................	53.6	7.1	15.7	22.6	1.0
Vegetal　　"　　..............	50.5	6.8	18.0	24.2	0.5
Animal (flesh) Fibrin.......	54.1	7.3	16.0	21.5	1.1
Vegetal (wheat)　"　.......	54.3	7.2	16.9	20.6	1.0
Animal (blood)　"　........	52.6	7.0	17.4	21.8	1.2

There is a corresponding likeness in the general properties and reactions of proteids. They are for the most part soluble ; they are colloidal or non-diffusible, i.e., they will not pass through the membrane of a dialyser, or only with great difficulty ; they are rarely crystalline ; they rotate the plane of polarized light to the left. Though not all soluble in water, they may be dissolved by the aid of heat in strong acetic acid and in caustic alkalies, but are insoluble in cold absolute alcohol and in ether. They may be precipitated from solution by strong mineral acids, etc., and their presence may be detected by methods more fully outlined in the practical study (p. 43). Many proteids are precipitated by heat (a process which is called *coagulation*) ; and it is worthy of note that temperatures which produce coagulation of proteids (40°–75° C.) produce also the death of organisms.*

---

* " Amongst the organic proximate principles which enter into the composition of the tissues and organs of living beings, those belonging to the class of *proteid* or *albuminous* bodies occupy quite a peculiar place and require an exceptional treatment, for they alone are never absent from the active living cells which we recognize as the primordial structures of animal and vegetable organisms. In the plant, whilst we recognize the wide distribution of such constituents as cellulose and chlorophyl, and acknowledge their remarkable physiological importance, we at the same time are forced to admit that they occupy altogether a different position from that of the proteids of the protoplasm out of which they were evolved. We may have a plant without chlorophyl, and a vegetable cell without a cellulose wall, but our very conception of a living, functionally active, cell, whether vegetable or animal, is necessarily associated with the integrity of its protoplasm, of which the invariable organic constituents are proteids.

" In the animal, the proteids claim even more strikingly our attention than in the vegetable, in that they form a very much larger proportion of the whole organism, and of each of its tissues and organs. We may indeed say that the material substratum of the animal organism is proteid, and that it is through

The molecular instability of proteids is proven by the ease with which they may be decomposed into simpler compounds; their complex constitution by the numerous compounds, themselves often highly complex, which may thus be derived or split off from them.

Amongst the other matters found in protoplasm those of most frequent occurrence and greatest physiological importance are two groups of less complex substances, viz., **carbohydrates** and **fats**. These contain carbon, hydrogen, and oxygen, but no nitrogen; they do not appear to be closely related to **proteids** in chemical constitution, but they occur to some extent almost everywhere in living organisms, and in many instances are known to be of great importance, especially in nutrition. They are rich in potential energy and mobile in molecular arrangement; hence it is not strange that they figure largely in food, and are often laid by as reserve food-materials in the organism.

Carbohydrates. These substances are so called because, besides carbon, they contain hydrogen and oxygen united in the same proportions as in water. They include starch, various kinds of sugar, cellulose, and glycogen. Starch ($C_6H_{10}O_5$) is readily converted into glucose or grape-sugar ($C_6H_{12}O_6$) in the tissues of plants and animals by the more or less direct influence of the protoplasm—a metamorphosis which, as will hereafter appear, is of paramount importance in the process of nutrition.

Fats. These are of especial importance as reserves of food-materials (e.g., in adipose tissue and in seeds). They contain much less oxygen than the carbohydrates; are therefore more oxidizable, and richer in potential energy.* They are insoluble in water,

---

the agency of structures essentially proteid in nature that the chemical and mechanical processes of the body are effected. It is true that the proteids are not the only organic constituents of the tissues and organs, and that there are others, present in minute quantities, which probably are almost as widely distributed, such as for instance phosphorus containing fatty bodies, and glycogen, yet avowedly we can (at the most) only say *probably*, and cannot, in reference to these, affirm that which we may confidently affirm of the proteids—that they are indispensable constituents of every living, active, animal tissue, and indissolubly connected with every manifestation of animal activity." (Gamgee, *Physiological Chemistry*, Chap. I.)

* According to recent researches, one pound of butter contains 5654 foot tons, and a pound of sugar 2755 foot tons, of energy. A pound of proteid is nearly equivalent in this respect to a pound of carbohydrate.

but dissolve readily in ether and chloroform. Treated with aqueous solutions of alkalies they are converted into soluble soaps.

Physical Relations. The appearance, consistency, etc., of protoplasm have already been described; but it still remains to speak of certain of its other physical properties, and especially of the manner in which its activity is conditioned by various physical agents.

Non-diffusibility. Living protoplasm, like most of the various proteid matters which it yields (p. 37), is *indiffusible.* It will be seen eventually that osmotic processes play a leading *rôle* in the lives of plants and animals, since they are in large part the means by which nutriment is conveyed to the living substance. In view of this fact, the non-diffusibility of protoplasm as well as of ordinary proteids is a fact of great significance.

Relations of vital action to temperature. It is a general law that within certain limits heat accelerates, and cold depresses, the activity of protoplasm. We know that cold tends to benumb our own bodies (provided they become really chilled), and in animals the heart beats more slowly, the movements become sluggish or cease, breathing becomes slow and heavy,—in a word, all of the vital actions become depressed,—whenever the ordinary temperature is sufficiently lowered. If we chill the rotating protoplasm of *Chara* or *Nitella*, the vibrating cilia of ciliated cells, or an actively flowing *Amœba*, the movements become slower, and finally cease altogether.

On the other hand, moderate warmth favors protoplasmic action. Benumbed fingers become once more nimble before the warmth of the fire. In a hot room the frog's heart beats more rapidly, cilia lash more energetically, the *Amœba* flows more rapidly, and the protoplasm of *Chara* courses more swiftly. In the winter months the protoplasm of plants and of many animals is in a state of comparative inactivity. Most plants lose their leaves and stop growing; many animals bury themselves in the mud or in burrows, and pass the winter in a deep sleep, during which the vital fires burn low and seem well-nigh extinguished. The warmth of spring re-establishes the activity of the protoplasm, and in consequence animals awake from their sleep and plants put forth their leaves.

But this law is true only within certain limits. Extreme heat and cold are alike inimical to life, and as the temperature approaches these extremes all forms of vital action gradually or suddenly cease. The limits are so variable, that it is not at present possible to formulate any exact law which shall include all known cases. For instance, many organisms are killed at the freezing-point of water ($0°$ C.); but certain forms of life have withstood a temperature of $-87°$ C. ($-123°$ F.), and recent experiments show that frogs and rabbits may be chilled to an unexpected degree without fatal results.

The upper limit is also inconstant, though less so than the lower. Most organisms are destroyed at the point of boiling water ($100°$ C.), but the spores of fungi have been exposed to a much higher temperature without destruction ($120°$–$125°$ C.). As a rule, protoplasm is killed by a temperature varying from $40°$ to $50°$ C., the immediate cause of death being apparently due to a sudden coagulation (p. 37) of certain substances in the protoplasm. Thus, if a brainless frog be gradually heated, death ensues at about $40°$ C., and the body becomes stiff and rigid (*rigor caloris*) from the coagulation of the muscle-substance. The lower forms of animal life agree well with plants in their "fatal temperatures," which in most cases lie between $40°$ and $50°$ C.

Lastly, it appears to be true that there is a certain most favorable or *optimum* temperature for the protoplasm of each species of plant and animal, this optimum differing considerably in different species. Probably the highest limit occurs among the birds, where the uniform temperature of the body may be as high as $40°$ C. The lowest occurs among the marine plants and animals of the Arctic seas, or of great depths, where the temperature seldom rises more than a degree or two above the freezing-point. Between these limits there appears to be great variation, but $35°$ C. may be taken as the average optimum.

Moisture. Protoplasm always contains a large amount of water, of which indeed the lifeless portion of living things chiefly consists. (See table on p. 35.) All plants and animals are killed by complete drying, though some of the simpler forms resist partial drying for a long time, becoming quiescent and reviving again when moistened, sometimes even after the lapse of years. Hence water appears to be an essential constituent of protoplasm, though, as in the case of mineral matters, we do not know the nature of its connection with the other elements or compounds present.

Electricity. It has been shown that many forms of vital action are accompanied by electrical disturbances in the protoplasm. It is therefore not surprising that the application of electricity to living protoplasm should have a marked effect on its actions. If the stimulus be very slight, protoplasmic movements are favored. Colorless blood-corpuscles creep more actively, and ciliary action increases in vigor. Stronger shocks cause a spasmodic contraction of the protoplasm (*tetanus*), from which it may or may not be able to recover, according to the strength of the shock.

Poisons. Towards certain agents protoplasm is indifferent or seemingly so, but towards others it behaves in a very remarkable manner. The matters known as poisons modify or destroy its activity, as is well known from the familiar effects of arsenic, opium, etc. Disease may also interfere with its normal activity; but the consideration of these phases of the subject belongs to the more exclusively medical sciences, such as toxicology and pathology.

Other physical agents. The more highly specialized forms of protoplasm are affected by a great variety of physical agents, such as light, sound, pressure, etc., and upon this susceptibility depend many of the higher manifestations of life. For instance, waves of light or of sound, acting upon special protoplasmic structures in the eye and ear, call forth actions which ultimately result in the sensations of sight and hearing. Similar considerations apply to the senses of smell, taste, and touch; but the discussion of all these special modes of protoplasmic action must be deferred. Enough has been said to show that living organisms (that is, the protoplasm which is their essential part) are able to respond to many influences proceeding from the world in which they live. Upon this property depend the intimate relations between the organism and its environment, and the power of adaptability to the environment, which is one of the most marvellous and characteristic properties of living things.

Vegetal and Animal Protoplasm. The protoplasm of plants is essentially identical with that of animals in chemical and physical relations, and manifests the same fundamental vital properties.

But it would be absurd to suppose this identity absolute, for if it were so, plants and animals would also be identical; and furthermore, the protoplasm of every species of plant and animal must differ more or less from the protoplasm of every other species. What is meant is, that the differences between the many kinds of protoplasm are secondary to the fundamental resemblances which underlie them.

PRACTICAL STUDY.

A. Protoplasm.

1. Take a long cell of *Nitella* or *Vaucheria*, lay it upon a slide, cut off one end, and with the back or handle of a scalpel press the cell gently from the other end so as to drive out the contents. Observe with the naked eye the small drop of viscid fluid, lift some of it upon a needle-point to test its viscidity, then cover, and examine with the highest power at hand. The colorless portions of the whole mass are often called *protoplasm*, but only the clearest portion, filled with innumerable fine black particles, is really entitled to the name. (If no long-celled plant is at hand, any soft cells, e.g., those of *Spirogyra* or yeast, may be crushed by pressure upon the cover-glass, and will answer.)

2. Add iodine, and note the effects.

3. Examine the protoplasm in cells of cartilage, in sections of the cotyledons of sprouting beans, in sections of the leaf-stalks of young leaves of geranium (*Pelargonium*), or in other objects furnished by the demonstrator, and report your results.

4. Study the movements of protoplasm in vegetal cells—e.g., *Chara, Nitella,* or *Tradescantia*—and in some animal cell—e.g., *Amœba, Gregarina* or colorless blood-corpuscles—or in other objects, and compare p. 26.

5. Snip off with fine scissors a bit of the gill of an oyster or a clam, mount it in normal fluid or in the blood of the animal, and examine with a high power. Find a spot where the *cilia* are moving slowly, and observe the ciliary action. Sketch.

6. In the same preparation employed in (5), or in a fresh one, note the effects of carmine finely rubbed up in a mortar with water and run under the cover-glass. Compare the result of this

experiment with another in which a carmine *solution*—e.g., Beale's carmine or borax-carmine—is used. Notice especially in the latter case the failure of the cells to stain until they are killed. How do you determine when the cells are dying and when they are killed?

7. Repeat (5) and (6), using ciliated cells scraped from the hinder part of a frog's mouth or from the œsophagus.

8. Lay open the œsophagus of a frog (or better, of a turtle) from the lower side, and pin it out flat with the inner surface exposed. Place bits of moistened cork on the surface at the anterior end. They immediately begin to travel backward along the œsophagus, being propelled by the cilia. Incline the animal, and note that the pieces of cork may be carried up a steep slope.

9. To some fresh, healthy yeast from an actively fermenting fluid add, under the cover-glass, an aqueous solution of eosine, magenta, or aniline blue. Notice that very few cells are immediately colored. Repeat the experiment, using yeast which has been boiled for a moment. Report the result to the demonstrator. Compare these experiments with a similar series in which the ordinary iodine solution is used instead of eosine, etc., and explain the difference in effect in the two cases.

B. Proteids.

1. Examine the "white" of an egg and cut it up with scissors. Add its own bulk of water, and filter. Put some of the filtrate into several small test-tubes, setting aside the rest for E. 2, and apply the following tests for proteids:

Tube No. 1. Add a few drops of strong nitric acid (HNO_3). A dense white or yellowish precipitate is formed. Boil for a moment, cool, and cautiously add ammonia (NH_4OH). The precipitate becomes of a bright orange color. (Xanthoproteic reaction.)

Tube No. 2. Add a few drops of caustic soda solution, then one or two drops of a solution of cupric sulphate, and heat. Note the violet color.

Tube No. 3. Heat in a water-bath, and note the coagulation.

Tube No. 4. Heat as in (3), but more slowly, and determine with a thermometer the coagulating point. Record the results in every case.

C. Carbohydrates.

1. Mount a little powdered starch (potato starch if possible) in water, examine with the microscope, and run in *dilute* iodine solution. The blue color thus produced is a characteristic reaction for *starch* ($C_6H_{10}O_5$).

2. Make a number of sections of some succulent vegetal stem; e.g., stem of *Pelargonium*. Place them for a few minutes in strong iodine solution, wash with water, and transfer to strong sulphuric acid (2 parts acid, 1 part water, by volume). Mount the sections in the acid, and examine. In successful experiments the cell-walls become blue—a characteristic reaction for *cellulose* ($C_6H_{10}O_5$).

3. Make in a test-tube a very dilute solution of glucose or grape-sugar ($C_6H_{12}O_6$); add a drop or two of a weak solution of cupric sulphate ($CuSO_4$), then several drops of a solution of sodium or potassium hydrate (KOH), and boil the fluid. A reddish precipitate of cuprous oxide is formed as follows:

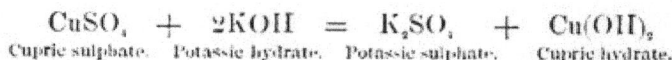

$$CuSO_4 \ + \ 2KOH \ = \ K_2SO_4 \ + \ Cu(OH)_2$$

Cupric sulphate.　Potassic hydrate.　Potassic sulphate.　Cupric hydrate.

On heating in the presence of glucose, the cupric hydrate is reduced to cuprous oxide (Cu_2O). This is known as "Trommer's test" for glucose.

D. Fats.

1. Shake up some olive oil with water in a test-tube. Note that they do not mix.

2. Repeat the experiment, using diluted white-of-egg instead of water. The fluid becomes milky white. Examine with the microscope, and observe that the oil is broken up into minute drops. The fat has been "*emulsified*," and the mixture is an *emulsion*.

3. Mount some milk (which is likewise an emulsion). Examine with the microscope, and compare with (2).

4. Add strong alcohol to some oil in a test-tube. They do not mix.

5. Heat the test-tube. The fluids mix completely. Upon cooling the oil separates and the fluid becomes turbid.

6. Add to some oil in a test-tube either turpentine, chloroform, or ether. Report your results.

E. Dialysis.

1. Place in a freshly prepared dialyser a strong (50%) solution of glucose, and put the dialyser in a vessel of distilled water. Set the vessel aside for half an hour. Note the rise of liquid in the dialyser. Then test the water outside the dialyser for glucose by Trommer's test. What do you conclude?

2. Place in a fresh dialyser some diluted white-of-egg. Place the dialyser in distilled water as before, and after a number of hours (5–24) test the water for proteids by one of the tests given above. What do you infer?

3. Test another portion of the water with silver nitrate solution, and another with barium chloride. What do you infer?

F. Temperature.

1. Observe the beating heart of a frog whose brain has been destroyed by the demonstrator. Count the beats at the ordinary temperature; again, after the organ is bathed with iced normal fluid; then when treated with normal fluid heated to 38° C.; and finally when treated with normal fluid at 65°. Record your results.

2. If possible, repeat this experiment with the heart of an embryo three-day chick, in which similar effects may be beautifully shown.

3. Mount cells of *Nitella, Anacharis,* or *Vallisneria,* showing rotating protoplasm, and support the ends of the slide by bits of thick pasteboard so that they do not come in contact with the stage of the microscope. Having noted the rate of flow, place a large heated coin on the slide near the object. If necessary, reheat and replace the coin until an effect is observed. Note the results.

4. Now replace the coin by one which has been in iced water, renewing if necessary. Note the results.

5. Heat some active *Nitella* thoroughly to 50° C. Examine with the microscope, and note the results.*

---

* These experiments can be more accurately performed by using a warm stage.

6. Place a sprig of *Anacharis*, having cut its stem squarely across, in a test-tube with fresh water (not distilled), and hold it in the sun-light. In successful experiments bubbles of oxygen come forth from the cut end at regular intervals. (It is necessary to have vigorous plants, and usually a number of sprigs must be tried.) These bubbles are produced by the chemical action of protoplasm, aided by chlorophyll (see p. 85). Count and record the number of bubbles per minute.

7. *Without changing the illumination*, immerse the base of the test-tube in iced water. Interpret and record your results.

8. Without moving the apparatus, gradually heat the water. Count and record the number of bubbles at different temperatures (e.g., 10°, 20°, 30°, 40°, 50°, C.). Note at what temperature the bubbles cease.

G. Rigor Caloris.

Immerse in a beaker of water at the ordinary temperature the hind limbs of a recently-killed frog. (Both brain and spinal cord should have been destroyed.) Heat the water gradually, and note by means of a thermometer the temperature at which the limbs stiffen, become rigid, and violently extended.

H. Rigor Mortis.

Examine an animal which has passed into the death-stiffening. This is regarded as a consequence of coagulation taking place at or near the time of the death of the muscles.

Compare with an animal in *rigor caloris* (G.).

Compare both with fresh living muscle from the leg of a decapitated frog.

I. Moisture.

1. Wipe dry with a towel or filter-paper a frog which has been killed by the demonstrator; then weigh, and place it in a warm bath made for drying at 100° C. If this be not at hand a hot-air register or radiator will answer. After a day or two weigh again, and calculate the loss of weight, actual and percent., caused by drying. This represents approximately the amount of water originally present.

2. Repeat the experiment, using instead of a frog a lettuce-head, a bunch of celery, a mushroom, a leaf, a potato, or any *fresh* and *succulent* vegetal structure. Compare the results with (I).

J. Poisons.

1. Note the convulsive twitchings of the muscles of a decapitated frog into which, a few moments previously, the demonstrator has injected subcutaneously a trace of *strychnine*. The convulsions become violent when stimuli are applied, and are caused by the unusual activity of the poisoned protoplasmic nervous structures.

2. Add to a fermenting yeast-fluid, which is giving off bubbles plentifully, a quantity of *corrosive sublimate* (a violent poison). The yeast is killed, as is proven by the stoppage of the fermentation.

K. Electricity.

1. If an interrupted current be sent into a muscle removed from a frog's leg, it will change shape or *contract*.

2. Compare the living frog's muscle in this respect with muscle in *rigor caloris* or *rigor mortis*. (G.) (H.)

CHAPTER IV.

THE CELL

CELLS were observed in plants more than two hundred years ago by the English botanist Robert Hooke, who described cork (a part of the bark of certain trees) as made up of " little boxes or cells distinct from one another." The word cell was afterwards employed by other observers to designate the minute crowded cavities appearing in sections of wood and other vegetal tissues, and the term soon came into general use.

The name itself shows that cells were at first regarded as cavities like the " cells" of a honeycomb, surrounded by solid walls. Even as late as 1840 many biologists were inclined to regard the walls as their essential part. But as biological knowledge advanced, it was shown that the walls might be wanting, and that the active and essential part of the cell is the living protoplasm, not the cavity in which it lies or the lifeless walls which surround it. The word cell became therefore as inappropriate as it would be if applied to the honey within the honeycomb. Nevertheless, by a curious conservatism, the term was and is retained to designate these structures, whether occurring in masses as constituent units of the plant or animal body, or leading independent lives as " one-celled" organisms.

It was first fully established by Schleiden in 1839, that higher plants are entirely built up of cells ; and about a year later Schwann demonstrated that animals have a like composition. This generalization is universally recognized as the " cell-theory" of Schleiden and Schwann, though it is no longer a theory, but simply a statement of fact. This conception of the structure of higher organisms was subsequently extended to their physiology, their vital actions being conceived as the sum of the co-ordinated individual actions (i.e., the lives) of the cells. For example, a muscle is not only composed structurally of cells, but its function of contraction (cf. p. 26) is likewise the resultant of the contrac-

tions of the cells. And so with other functions and the life of
the organism generally.

The present chapter will deal chiefly with the form and structural relations of cells.

A Typical Cell. The simplest way to understand cells and
their modifications is to study their development. The germ
(egg, spore, etc.) from which every higher plant or animal arises
is at first a single cell, called the *germ-cell*, which forms part of
the body of the parent and
is essentially similar to other
cells. Sooner or later the
germ-cell is detached from
the parent, and gives rise
by a process of multiplication and growth to a mass
of cells which constitutes a
new individual, like the
parent. The structure of
such a germ-cell is well
illustrated by the egg of a
star-fish (Fig. 20). The egg
of every animal, even in-
cluding man, shows the same

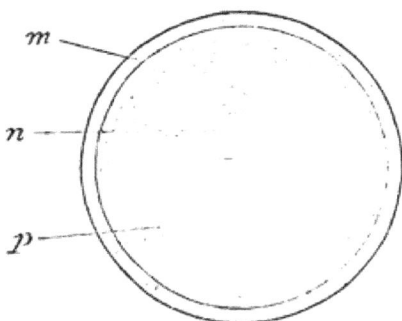

Fig. 20.— Slightly diagrammatic figure of the
egg or ovum of a star-fish, showing the structure of a typical cell; *m*, membrane; *n*, nucleus; *p*, protoplasm.

essential structure: that of the star-fish shows it with especial
clearness, and this may be taken as typical, not only of all eggs,
spores, etc., but also of all cells.

Form. Like most free or isolated cells, the egg is nearly
spherical. In the tissues, cells usually have angular contours as a
result of mutual pressure—just as soap-bubbles, which are spherical when free, become polyhedral and bounded by plane surfaces
when a number of them are brought in contact. The sphere is
therefore regarded as representing the typical form of the cell;
and from this all other forms are derived.

Structure. In such an egg, as in any other typical cell, three
parts may be distinguished. These are (1) cell substance, (2)
nucleus, (3) membrane (or cell-wall).

In perfectly typical cells the *cell-substance* is simply a mass
of unmodified protoplasm. In the star-fish ovum it contains also
scattered granules of lifeless proteid and fatty substances (here

4

called *deutoplasm*), which serve as a store of food for the developing embryo.

The *nucleus* is a rounded body suspended in the cell-substance; it is distinguishable from the latter by its higher refractive power, and by the intense color it assumes when treated with staining fluids. It is surrounded by a very thin membrane, and consists internally of a clear fluid (*achromatin*), through which extends an irregular network of fibres (*chromatin*). It is especially these fibres which are stained by dyes. In the meshes of

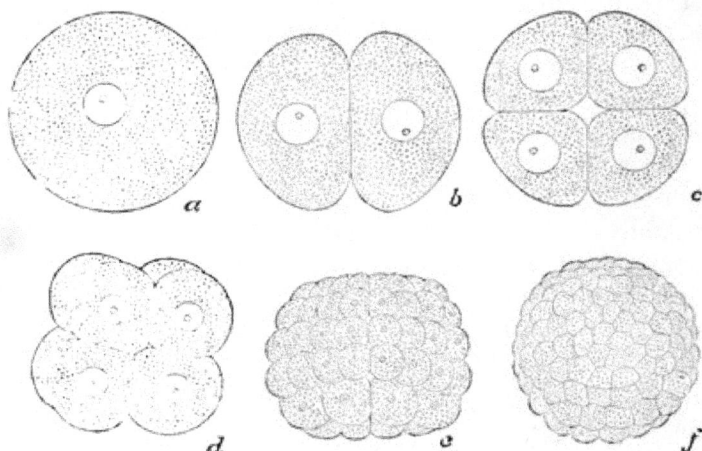

FIG. 21.—Cleavage or segmentation of an ovum, showing successive division of the germ-cell (*a*) into two (*b*), four (*c*), and eight (*d*). Later stages are shown at *e* and *f*. The first four figures are diagrammatic; *e* and *f* are after Hatschek's figures of the development of a very simple vertebrate (*Amphioxus*).

the network is suspended a second rounded body known as the *nucleolus*, which stains even more deeply than the network itself.

The *membrane* or *wall* of the cell forms a rather thick sac, composed of a soft, lifeless material closely surrounding the cell-substance.*

Cells in the Body. Typical cells, closely similar to that just described, are found here and there throughout the vegetal and animal kingdoms in the most unlike organisms. Such cells show

---

* The word cell has been used in Chap. I. and elsewhere to denote the living matter within the membrane, the latter being considered a product of the cell rather than an integral part of it. It is more usual to include the membrane in a definition of the cell, and as a matter of convenience it is so included here.

us the model after which all others are built, in spite of endless variations in detail; for the typical germ-cell gives rise to all other cells in the body. This process is as follows:—

The first step of development (Fig. 21) consists in the division of the egg into two similar halves, which differ from the original cell only in lacking membranes, both being surrounded by the membrane of the original cell. Each of the halves divides into two, making four in all; these again into two, making eight, and so on throughout the earlier part of the development. By this process (known as the cleavage or *segmentation* of the egg) the germ-cell gives rise successively to 2, 4, 8, 16, 32, 64, etc., descendants, forming a mass of nearly similar cells, out of which, by still further division and growth, the body of the future animal is to be built up. These cells are only slightly modified, but differ in most animals from the typical germ-cell in having at first no surrounding membranes. The membrane of the original germ-cell meanwhile disappears.

The embryos of all higher plants and animals are derived from the germ-cell by a process essentially like that just described, though both the form of the cells and the order of division is usually more or less irregular. In animals the cells thus formed are usually naked at first, though they often acquire membranes in later stages. Among plants, on the contrary, the cells usually possess membranes from the first, probably because their need for a firm outer support is greater than the need for free movement demanded by animal cells.

Modification of the Cell. Differentiation. The close similarity of the embryonic cells does not long persist. As development proceeds the cells, continually increasing in number by division, become modified in different ways, or *differentiated*, to fit them for the many different kinds of work which they have to do. Those which are to become muscle-cells gradually assume an entirely different form and structure from those which are to become skin-cells; and the future nerve or gland cells take on still other forms and structures. The embryonic cells are gradually converted into the elements of the different tissues—this process being the *differentiation of the tissues* which has already been mentioned on p. 15—and are in this way enabled to effect a *physiological division of labor.*

The variations in form and structure which thus appear are

endlessly diversified. Cells may assume almost any conceivable form, and there are even cells (e.g., *Amœba*, or the colorless corpuscles of the blood) which continually change their form from moment to moment, passing through grotesque transformations which would be incredible did we not actually behold them. The variations in structure may involve any or all of the three characteristic parts of the typical cell, being at the same time accompanied by variations of form. It is easy to understand therefore, how cells may vary endlessly in appearance, while conforming more or less closely to the same general type.

Variations in Form. Free cells are, as a rule, bounded by curved surfaces (as in the star-fish's ovum), while those which are united in tissues, especially when crowded, are

FIG. 22.—Series of cells showing gradations between spherical and elongated forms. (From various sources and differently magnified.)

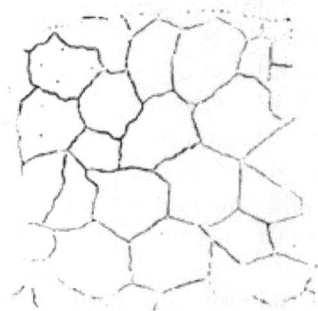

FIG. 23. (After Klein.)—Flattened cells arranged in a layer ("pavement epithelium"), from the surface of the diaphragm of a rabbit.

usually bounded by plane surfaces and possess angular contours, owing to mutual pressure. If yielding spherical bodies (or cells) be crowded together they tend to assume regular twelve-sided forms (dodecahedrons), as may be seen in the experiment with soap-bubbles already referred to.

Now in comparing the forms of cells we may trace every gradation from the typical spherical cell, taken as a starting-point, (or its polyhedral representative among angular cells) through oval, oblong, cylindrical, or prismatic forms, to those which are drawn out into attenuated filaments and may be hundreds of times longer than their width. Cells which are considerably elongated are often called *fibres;* thus we speak of muscle-fibres,

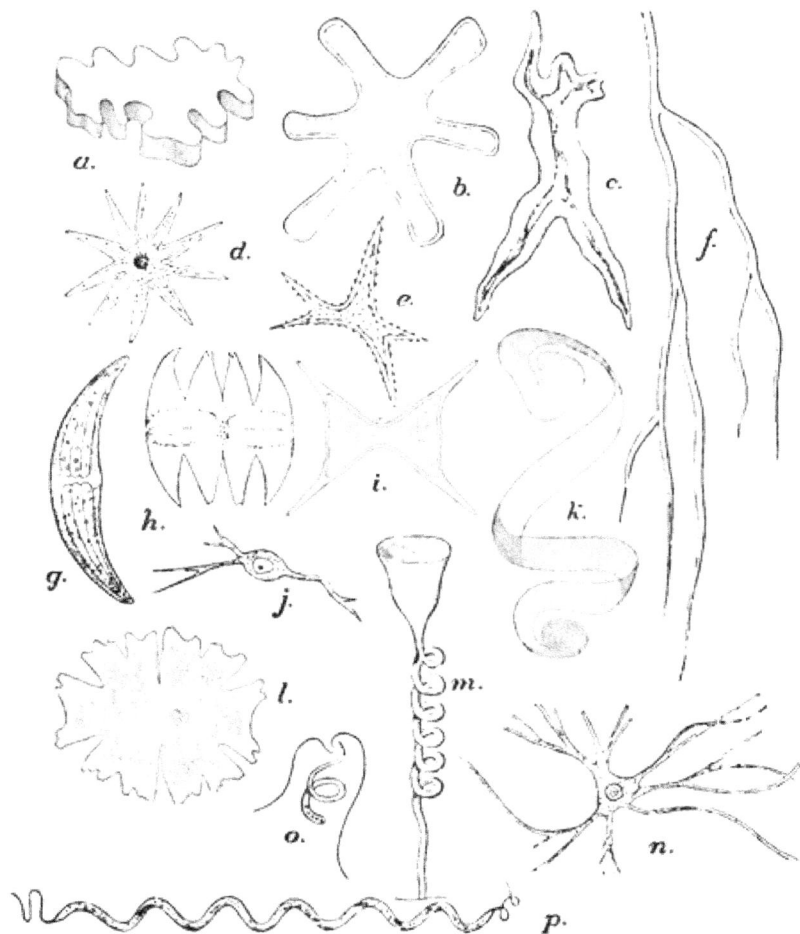

Fig. 24.—A few examples of the infinite variety of cell-shapes, etc. *a*, epidermal cell of *Callitriche ; b*, stellate cell of the flowering rush, *Butomus ; c*, branched bast cell of the larch *; d*, stellate hair-cell of *Deutzia ; e*, stellate cell of the yellow water lily, *Nuphar; f*, part of a branching laticiferous cell of *Euphorbia ; g*, a one-celled plant, *Cosmarium; h*, a one-celled plant. *Micrasterias ; i*, a one-celled plant *Staurastrum ; j*, a cartilage cell from the human metatarsal bone *; k*, the outer cell wall of a pollen grain of *Thunbergia* normally wound about the (spherical) grain, but now unrolled by reagents *; l*, another form of *Micrasterias ; m*, diagram of an infusorian animal, *Vorticella ; n*, nerve-cell from the spinal cord of a calf *; o*, germ-cell (spermatozoid) of a stone-wort, *Nitella ; p*, a one-celled plant or microbe, *Spirillum*. (Compiled from various sources, and not equally magnified.)

nerve-fibres, etc., though fibres are not always cells. A few examples illustrating such a series are shown in Fig. 22. Towards their ends elongated cells may be angular, rounded, attenuated, or branched. If the cells are extended in two directions instead of one, flattened or scale-like forms are produced, which are often joined by their edges into a tissue resembling a pavement (Fig. 23). Seen from their flat surfaces, such cells may have the most diversified outlines—angular, wavy, toothed, irregularly branching, etc.—as shown in Fig. 24, *a*. Whether elongated, flattened, or approximately equilateral, cells may be produced into various projections, which give rise to such a diversity of forms that a volume would be required to describe them. These projections may be few and inconspicuous; but from this condition there is every possible gradation leading to cells which display innumerable lobes, branches, or filaments, and these may even be many times longer than the main body of the cell, sometimes forming the most complicated networks, or systems of branches. A very few of these different forms are shown in Fig. 24. For other examples the student is referred to the various figures throughout this book.

Variations in Structure. The variations in structure may affect the cell-substance, the membrane, or the nucleus; and in many cases one or more of these parts may be entirely wanting. The membrane is frequently absent during a part or the whole of the life of the cell. On the other hand, the cell-substance and nucleus often disappear from old cells, for example in the case of wood (Fig. 3), leaving a cavity filled with air or water. Such cells scarcely merit the name, being really the dead remains of cells; though it is a curious fact that the word *cell* was first applied to precisely such structures (p. 48). It is still an open question, whether the nucleus is ever really wanting in living cells, though it is often very difficult to see, and its existence has been denied in a number of the lower forms of life.

It is hardly worth while to make at this point more than a brief mention of the manifold special variations in the structure of cells, since these will be better understood in the course of future study.

Cell-substance. In the last chapter some account has been given of the deposition of lifeless matter and of other changes taking place in the cell-substance. Most of these modifications need not be further dwelt upon at present, but one of them should be briefly considered. It has been stated that drops of liquid often appear in the protoplasm—e.g., sap, oil, etc. These drops are sometimes single, in which case the drop often becomes so large as to occupy the entire central part of the cell, the protoplasm becoming reduced to a layer (sometimes extremely thin) which lines the cell-membrane (cf. *Nitella*, p. 28), and is known as the *primordial utricle* (see p. 28). In other cases a number of drops appear in the same cell. If these increase greatly in size, as they often do, they may occupy the greater part of the cell, the protoplasm being reduced to a more or less fine network which connects with a peripheral layer (cf. *Tradescantia*, p. 30). The cavities in which the drops lie are called *vacuoles*, and the cells are said to be *vacuolated*.

Nucleus. The form of the nucleus is tolerably constant, being usually round or oval; but sometimes it may be irregular or elongated. It may contain one or more nucleoli, and the character of the internal network (p. 50) varies extremely. During the division of cells (as in the developing egg of the starfish) the nucleus undergoes marvellous transformations (known as *karyokinesis* or *mitosis*) which are still imperfectly understood, and are the object of active research at the present time.

Cell-membrane. The variations of the membrane demand a somewhat fuller notice at this point. Wholly wanting in many young cells, it first appears as a delicate film outside the cell-substance, and may continue to increase until its thickness is enormous (Fig. 4). Chemically considered, it differs in plants and animals. In plants it is usually composed of a carbohydrate substance known as cellulose $(C_6H_{10}O_5)$, which is closely related to starch. In animals it is composed of nitrogenous substances of unknown composition. As the cell grows older the membrane may undergo

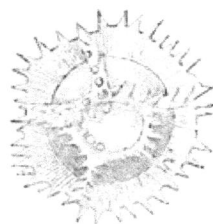

FIG. 25. (After Sachs) Pollen-grain (a single isolated cell of *Cobaea*), showing external prominences of the cell membrane.

many changes, both in configuration and in chemical composition. Various mineral deposits may take place in its substance, or on the surface; or its substance may undergo a direct transformation into wood (*lignification*), horn, gelatine, mucus, and any one of a great variety of other substances. For instance, in the stalks of many kinds of grasses and grains the outer cell-walls become impregnated with solid deposits of silica, sometimes to such an extent that the exact form of the cell is retained when the tissue has been burned or the soft parts have been removed by reagents. These siliceous deposits perhaps give to the stalks the rigidity which enables them to maintain an erect position against wind and showers. Siliceous deposits likewise occur in the cell-walls of those microscopic one-celled plants known as

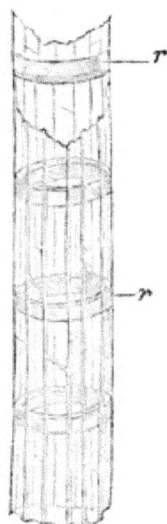

Fig. 26. (After Sachs.)—Part of a long cylindrical cell ("annular vessel") from the woody tissue of maize, showing ring-like or "annular" thickenings (r) on the inner surface of the membrane.

diatoms. So firm and enduring are the deposits, that they are nearly indestructible by most chemical and physical reagents. In certain parts of the world great beds of earth occur which consist almost entirely of the siliceous remains of diatoms of bygone geologic ages. These are so perfectly preserved that the various species can be recognized as clearly as in the case of diatoms now living. Calcareous deposits occur likewise in the cell-walls of many other plants, and may be so extensive (as in the *corallines*) that the whole plant appears hard and white, as if sculptured out of marble.

Among animals a similar process is common; examples are the shells of snails, oysters, and of other mollusks, the skeletons of corals, and the calcareous and siliceous coverings of a multitude of minute animals. These live for the most part in the sea or in fresh water, from which dissolved mineral matters are taken and laid down in the solid form. As in the case of diatoms, large beds of mineral matter (*e.g.*, *chalk*) have been formed by the accumulated remains of these animals. Indeed the recovery of solids from solutions is one of the most important geological effects produced by living things.

In the last place, the cell-membrane may undergo great changes in physical as well as chemical structure. These modifications are especially conspicuous and varied among plants. They consist for the most part in the appearance of prominences or depressions on either the outer or inner surface of the membrane.

or on both. The prominences appear as knobs, spines, hairs, ridges, etc., which may be quite irregular or may show definite arrangement (Fig. 25). In the elongated cells of certain parts of wood the prominences have the form of ridges, which project from the inner face of the membrane as rings, spiral bands, or reticular networks, as shown in Figs. 26, 27. A like diversity exists in the form and arrangement of the depressions in the cell-membrane. They often appear as rounded pits, which may extend

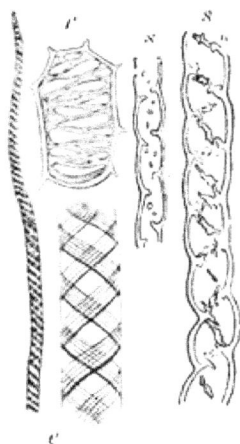

Fig. 27. (After Sachs.)—Cells from a liverwort (*Marchantia*), showing thickenings on the inner face of the membrane; *s, s*, rhizoids (i.e., root-like structures) with the thickenings in the form of irregular spiral bands; *r*, cell from the interior of the plant with irregular net-like ("reticular") thickenings; *e*, peculiar elongated cells (elaters) from the spore-cases, showing very regular spiral thickenings.

Fig. 28. (After Sachs.)—Elongated wood-cells ("pitted tracheids") from ordinary pine wood, showing rounded pits and holes in the membranes.

completely through the cell-membrane so as to form actual perforations or pores (Fig. 28). In very thick-walled cells (see Fig. 4) such pores form branching canals running through the walls. The pits or pores may be elongated so as to become oval or slit-like, as shown in Fig. 29, and in this case may be either transverse or oblique to the long axis of the cell.

In short, there is scarcely a limit to the variations in number,

form, and arrangement of the prominences or depressions which
may occur in cell-membranes. Many of them will become familiar in the course of future study.

Cell-membranes in Tissue. In tissues the membranes, if present, appear as partitions separating the cells. In young tissues,
the partitions usually appear single (Fig. 1); but as the tissue
grows older the partitions split along the middle so as to form two
layers, each of which may be regarded as the membrane of the

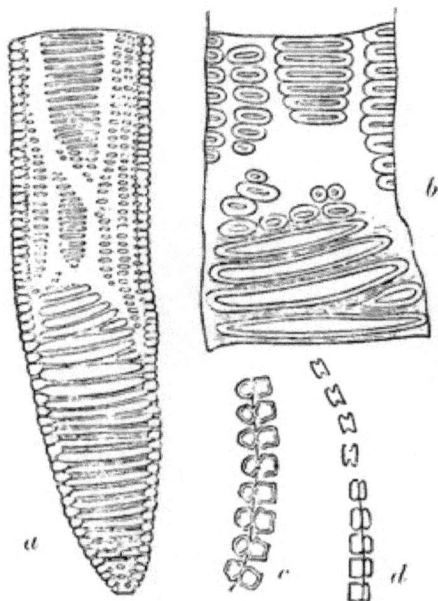

Fig. 29. (After De Bary.)—Elongated cells (ladder-cells, or "*scalariform tracheids*"), from
the underground stem of a fern (*Pteris aquilina*), having a series of transverse pits and
slits through the walls; *a*, part of a cell seen *en face*; *b*, membrane more highly magnified; *c*, section of the wall before the pits have broken through; *d*, shows how the pits
become slits.

cell which it surrounds (Fig. 2). The splitting of the partitions
is sometimes incomplete or does not take place at all. In the
latter case no trace of the double nature of the partitions can be
seen. If the separation is incomplete (as in Figs. 2 and 9), angular
intervals known as *intercellular spaces* are often left between
the cells. There is every stage between a complete independence
of the membranes and their complete fusion. But whether
separation be complete or not the partition-wall lying between

two adjacent cells is always to be regarded as a derivative of both (p. 16).

The partitions are often perforated by holes or canals, by means of which a free communication is established between adjoining cells. (See Fig. 4.) It has recently been demonstrated that such perforations often give passage to threads of protoplasm connecting the adjacent cell-bodies. The latter, in fact, are seldom or never completely isolated by the partitions which surround them. In some tissues the partition-walls may be partly or wholly wanting, the cell-bodies being in immediate contact.

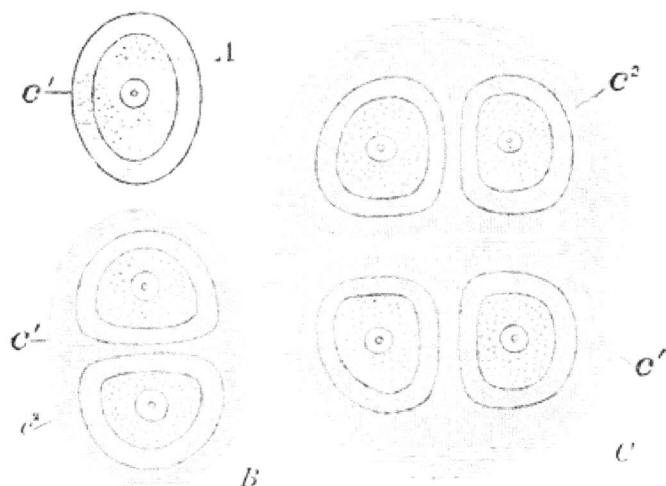

FIG. 30.—Diagram of developing cartilage: A, single cell surrounded by its membrane or capsule (c¹); B, the same cell divided into two, each of which is surrounded by a secondary capsule (c²); in C, the two have divided into four, each surrounded by a tertiary capsule; the primary and secondary capsules now form the matrix.

In other cases the cell-bodies are not only in contact, but actually fuse together, so that the whole tissue consists of a continuous mass of protoplasm, in which nuclei are imbedded. In this case the tissue is called a *syncytium*. It is exemplified by ordinary striped muscle, which consists of elongated fibres, each of which contains a number of nuclei and therefore represents a corresponding number of cells fused together.

It often happens, especially in the tissues of animals, that the partition-walls become of great thickness, the membranes fusing more or less completely together to form what is known as the intercellular substance or *matrix*. The cells then are scattered at intervals throughout the matrix. In some such cases the cells possess distinct individual membranes in addition to the matrix— or rather the cell-membranes consist of two layers, the inner being distinct, and the outer being fused with the corresponding layer of other cell-membranes to form the matrix; and finally, after or during the formation of the matrix, various deposits, often mineral, may take place within it (p. 18).

The facts just described are illustrated by the development of cartilage and bone. Young cartilage (as well as certain forms of the adult tissue) consists of rounded cells, surrounded by thin membranes, called the *capsules*, composed of a soft translucent substance (Fig. 30, *A*). As the tissue develops the cells continually multiply by successive division into two, four, eight, etc., and at each division the daughter-cells surround themselves with new capsules within the original one. The older capsules increase in thickness and fuse more or less completely together to form the matrix, the last-formed capsules alone remaining distinct. In many kinds of cartilage the lines of fusion between the capsules disappear, though in some cases they can be brought into view by the use of suitable reagents. The process just described is illustrated by Fig. 30, which represents the origin of four cartilage-cells from a single mother-cell, and shows the primary, secondary, and tertiary capsules.

Many bones consist at first of cartilage like that shown in Figs. 6 and 30. Solid calcareous salts are afterwards deposited at various points in the matrix (simultaneously with other changes which need not be mentioned here). These deposits increase in number, and ultimately fuse into a hard continuous mass. The cells at the same time undergo various modifications, and finally assume the branching form which is characteristic of them in the adult bony tissue (Fig. 8).

Summary. The ultimate units or cells of the body are constructed upon the same plan, sometimes appearing almost without modification in those known as typical, unmodified, or undifferentiated cells. Such cells consist of three parts, viz., cell-substance, nucleus, and membrane. Modifications or differentiations of the cell affect either its form or structure, or both, and may involve extensive chemical as well as physical changes. Modification of the cell-substance consists in the deposit within it of various lifeless substances; or of chemical or physical alteration of the protoplasm itself. In old cells the cell-substance may entirely disappear. The nucleus varies in form, size, internal structure, and physical consistency. It may, apparently, be wanting. The membrane is often wanting; when present it varies in chemical composition, thickness, and form, and is often complicated by the presence of various kinds of prominences, depressions, or perforations, variously arranged. In tissues, the individual membranes are either distinct or more or less fused together, and various lifeless matters may be deposited within their substance.

PRACTICAL STUDY.

Since the student has already examined a number of different cells, it is here only desirable to study a few additional forms to illustrate certain special points; and if the material cannot be procured the practical study may be omitted, wholly or in part.

1. Examine the small transparent ovum of some animal, either fresh or in the prepared specimen. The most favorable are those of starfish or sea-urchins, but those of earthworms, leeches, snails, or fresh-water mussels will answer. In the lack of these, examine prepared sections of the ovary of a mammal (cat, rabbit, etc.), and observe the ova. Fresh specimens should be treated with 2% acetic acid.

2. Observe some of the stages in the cleavage of the ovum. Specimens may usually be obtained by keeping various kinds of fresh-water snails (*Physa, Planorbis*, etc.) in aquaria, where they readily lay eggs. It is difficult to follow more than the first two or three divisions. In spring the process may be easily studied in the ova of toads, frogs, or salamanders.

3. Spread on a clean cover-glass a thin film of newt's or frog's blood, and dry perfectly in the air. Pour on a strong solution of hæmatoxylin (Grenacher's or Kleinenberg's), and leave for 10–15 minutes without allowing it to dry. Wash off the dye by a gentle stream of water from a faucet or poured from a beaker, add a very dilute aqueous solution of eosin; leave for a minute, wash again thoroughly, and then dry the cover. Add a drop of dis-

solved Canada balsam, invert the cover, and place in the centre of a slide. Examine with the microscope. In successful preparations the cells (*corpuscles*) retain their form nearly unchanged, the nuclei are colored bright blue, and the protoplasm (*stroma*) bright red.

4. Tease out in water the very young anther of some flowering plant preserved in alcohol. Examine, and treat with iodine. Treat another specimen with 2% acetic acid colored vivid green by the addition of a little aqueous solution of methyl green. Observe the protoplasm, cell-membranes, and nuclei. If the specimen be of the proper age, the cells (*mother-cells of the pollen-grains*) may be observed dividing to form the pollen-grains.

5. Examine mature pollen-grains of various kinds of flowering plants. Mount them dry on a slide.

6. Examine young spores from the developing fruit (*sporogonium*) of a moss or fern. Stain with iodine. Afterwards examine mature spores from another specimen.

7. Soak a piece of pine-wood in glycerine, make a delicate longitudinal section with a razor, mount in glycerine, and examine. Note the thick-walled, empty, elongated wood-cells, and the numerous openings (*bordered pits*) in the walls.

8. Make a thin longitudinal section radially through a well-developed stem of *Pelargonium*. Search for and examine—

    (*a*) Elongated cells (*pitted tracheids*) with walls marked by numerous hollows or pits;

    (*b*) Elongated cells (*spiral and annular vessels*) whose walls have spiral or ring-shaped thickenings inside.

9. Make a transverse section through the stem of the yellow water-lily (*Nuphar*). Examine the curious star-shaped cells scattered through the tissue. The stem of the white water-lily (*Nymphæa*) shows similar cells.

10. Examine hairs from the lower side of a leaf of *Deutzia*, which may be procured in any hot-house.

11. Examine collenchyma (tissue with the cell-membranes thickened at the angles) in cross-sections of the stem of *Pelargonium*, or Sunflower, or the leaf-stalk of *Begonia*.

12. Examine the sub-epidermal cells in cross-sections through

the midrib of *young* fern leaves.   Note the abundant protoplasm, large nuclei, central vacuole, and greatly thickened cell-walls.

We have now considered the general arrangement of living matter in organisms, and some of its more important properties. But it is impossible to derive adequate knowledge of vital structures and activities without prolonged and thorough study of complete individual organisms which may be taken as more or less typical of all.   In this way only can we acquire complete, connected and correct ideas of what living things are, how they come into being, how they are constructed, how they live, and how they are related to one another and to the lifeless world by which they are surrounded.   To this end a detailed account of a common plant will now be given, followed by a similar analysis of a common animal.   If the biology of these organisms be really mastered by studying the *specimens*, the descriptions and the figures, the student will have been fairly introduced to many of the most important principles of biology.

CHAPTER V.

THE BIOLOGY OF A PLANT.

THE COMMON BRAKE OR FERN (*Pteris aquilina*, Linnæus).

For the study of a representative vegetal organism some plant should be chosen which may be readily procured and is neither very high nor very low in the scale of organization. Such a plant is a common fern.

Ferns grow generally in damp and shady places, though they are by no means confined to such localities. Some of the more hardy species prefer dry rocks or even bold cliffs, in the crevices of which they find support; others live in open fields or forests, and still others on sandy hillsides. In the northern United States there are altogether some fifty species of wild ferns, but those which are common in any particular locality are seldom more than a score in number. Throughout the whole world some four thousand species of ferns are known, but by far the greater number are found only in tropical regions, where the climate is best suited to their wants. At an earlier period of the earth's history ferns attained a great size, and formed a conspicuous and important feature of the vegetation. At present, however, they are for the most part only a few feet in height. Nearly all are perennial; that is, they may live for an indefinite number of years. Most of them have creeping or subterranean stems; but some of the tropical species have erect, aerial stems, sometimes rising to a height of fifty feet or more and forming a trunk which is cylindrical, of equal diameter throughout, and bears leaves only at the summit, like a palm.

Of all the ferns perhaps the commonest and most widely distributed is the "brake" or "eagle-fern," which is known to botanists as *Pteris aquilina*, Linnaeus, or *Pteridium aquilinum*, Kuhn. This plant is not only common, but of comparatively simple structure; it is of a convenient size, and has been much studied. It may therefore be taken both as a representative fern and as a representative plant.

Habitat, Name, etc. The brake occurs widely distributed in the United States, under a great variety of conditions; e.g., in loose pine-groves, especially in sandy regions; in open woodlands amongst the other undergrowth; on hillside pastures and in thickets—indeed almost everywhere, except in very wet or very dry places. It appears to be equally common elsewhere; for, according to Sir W. J. Hooker, *Pteris aquilina* grows "all round the world, both within the tropics and in the north and south temperate zones. . . . . In Lapland it just passes within the Arctic circle, ascending in Scotland to 2000 feet, in the Cameroon Mountains to 7000 feet, in Abyssinia to 8000 feet or 9000 feet, in the Himalayas to about 8000 feet."—(*Synopsis Filicum.*)

" Pteris (πτέρις, the common Greek name for *fern*), signifying wing or feather, well accords with the appearance of *Pteris aquilina*, the most common and most generally distributed of European ferns. It is possible that this fern may rank as the most universally distributed of all vegetable productions, extending its dominion from west to east over both continents and islands in a zone reaching from Northern Europe and Siberia to New Zealand, where it is represented by, and perhaps identical with, the well-known *Pteris esculenta*. The rhizome of our plant like that of the latter is edible, and though not employed in Great Britain as food, powdered and mixed with a small quantity of barley-meal it is made into a kind of gruel called *gofio*, in use among the poorer inhabitants of the Canary Islands."—(Sowerby.)

The specific name *aquilina* (*aquila*, eagle) and a popular name, "eagle-fern," in Germany, etc., have come from a fanciful likeness of the dark tissue seen in a transverse section of the leaf-stalk, to the figure of an outspread eagle. The same figure has, however, been compared to an oak-tree, and has also given rise to the name of "devil's-foot fern," from its alleged resemblance to "the impression of the deil's foot," etc., etc.

The popular designation of this plant as "the brake" testifies to its great abundance; for a brake is a dense thicket or undergrowth—as for example a cane "brake."

When fully grown (Fig. 31) the common brake has a leafy top supported by a polished, dark-colored, erect stem, which in New England rises to a height of from one to four feet above

5

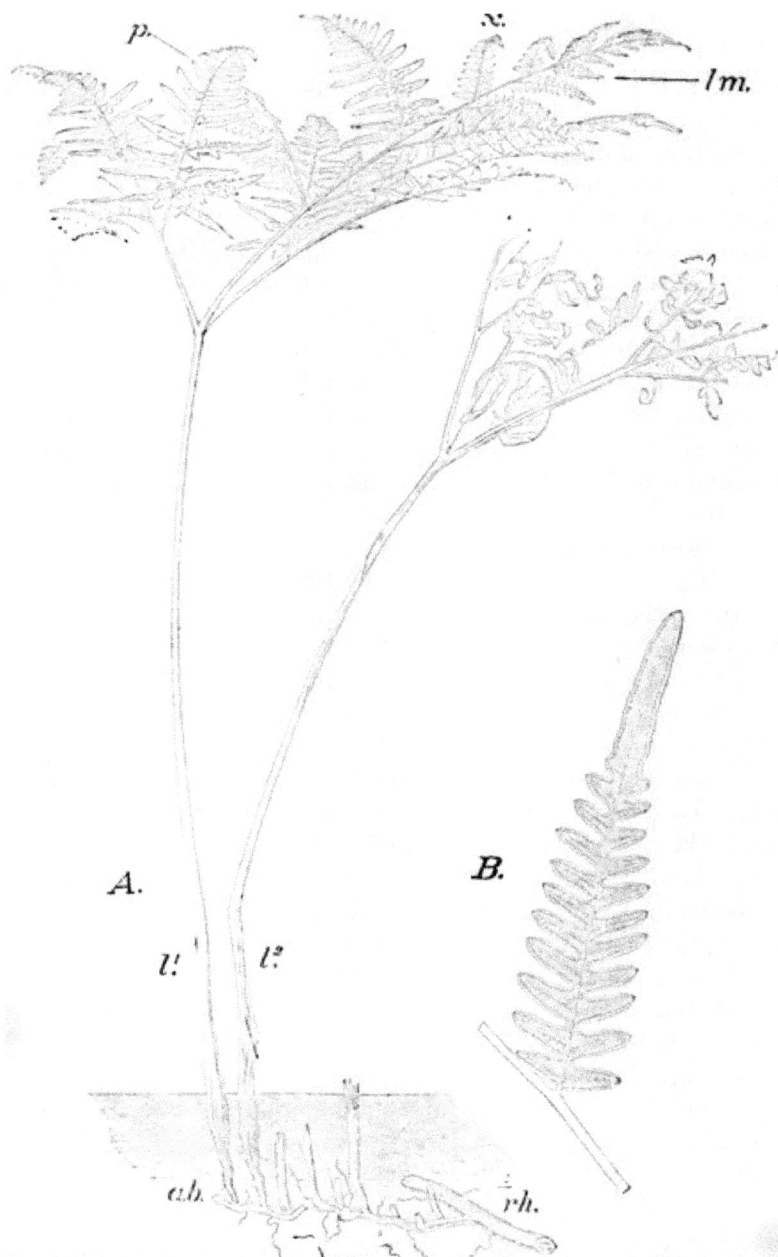

FIG. 31.—The Brake (*Pteris aquilina*), showing part of the underground stem and two leaves, one, (*l¹*), of the present year, in full development; the other, (*l²*), of the past year, dead and withered. *a.b.* apical bud at the extremity of a branch which bears the stumps of leaves of preceding years and numerous roots; *l¹*, mature active leaf; *l²*, dead leaf of preceding year; *lm*, lamina of leaf; *p*, pinna; *rh*, portion of main rhizome; *x*, younger pinna, which is shown enlarged at *B*. This pinna is nearly similar to the pinnules of older pinnæ. (× ½.)

the ground. In this climate, however, it appears to be somewhat
undersized, for it grows to a height of fourteen feet in the Andes,*
and in Australia attains to twice the height of a man, forming
a dense undergrowth beneath tree-ferns 40–100 feet high.† In
Great Britain it is from six inches to nine feet high (Sowerby),
or even larger in exceptional cases. "In dry gravel it is usually
present, but of small size; while in thick shady woods having a
moist and rich soil it attains an enormous size, and may often be
seen climbing up, as it were, among the lower branches and
underwood, resting its delicate pinnules on the little twigs, and
hanging gracefully over them."—(Newman.)

General Morphology.

The brake consists of two very different portions—one green
and leaf-like, which rises above the ground; the other black and
root-like, lying buried in the soil. These will henceforth be
spoken of as the *aerial* and the *underground parts*. The latter
lies at a depth of an inch to a foot below the surface, branching
widely in various directions. It may often be followed for a long
distance, and in such cases reveals a surprisingly complicated
system of underground branches.

Superficially, the underground portion of the fern appears to
be a root, but a closer examination shows it to be really the *stem*
or axis of the plant, which differs from ordinary stems only in
the fact that it lies horizontally under the ground instead of rising vertically above it. The aerial portion, which is often taken
for stem and leaf, is really leaf only. And the true roots are the
fine fibres which spring in great abundance from the underground
stem. Underground stems more or less like that of *Pteris*
are not uncommon—occurring, for instance, in the potato, the
Solomon's-seal, the onion, etc. In *Pteris* the stem is technically
called the *rootstock* or *rhizome*, and in this plant it constitutes the
larger and more persistent part of the organism. In the specimen
shown in Fig. 32 the rhizome was about eight feet long and bore
two leaves. It was dug out of sandy soil on the edge of a wood-
land, and lay from one to six inches below the surface. It was

---

* Hooker, *l. c.*
† Krone, *Botan. Jahresbericht*, 1876 (1), 346.

crossed and recrossed in all directions, both above and below, by the rhizomes of its neighbors, the whole constituting a coarse network of underground stems loosely filling the upper layer of the soil.

The *aerial part* (the *frond* or *leaf*) is likewise divisible into a number of parts, comprising in the first place the leaf-stalk or *stipe*, and the leaf proper or *lamina*. The latter is subdivided like a feather (*pinnately*) into a number of lobes (*pinnæ*, Fig. 32), which vary in form according to the state of development of the leaf. In large leaves the two lower pinnæ are often larger than the others, so that the leaf appears to consist of three principal divisions, and is said to be "*ternate*" or triply divided (Fig. 31, *A*). Each pinna is in turn pinnately subdivided into *pinnules* (*pinnula*) or leaflets (Fig. 31, B), each of which is traversed down the middle by a thickened ridge or rod, the *midrib*. The leaflets sometimes have smooth outlines, but are usually lobed along the sides, as in Fig. 31, B. In this case their form is said to be *pinnatifid*. Each lobe is likewise furnished with a midrib. The *stipe* enlarges somewhat just below the surface of the ground, then grows smaller and joins the rhizome. The

FIG. 32.—An entire plant of *Pteris*. One of the leaves is young and small, and a comparison of the figure with Fig. 31 will show some of the differences between leaves of different ages.

enlargement is of considerable interest, for it occurs at precisely the point of greatest strain when the leaf is bent by the wind or otherwise, and must serve to strengthen the stipe.

The plant as a whole may be regarded as consisting of an *axis* (the rhizome and its branches) which bears a number of *appendages* in the form of roots and leaves. The axis forms the central body or trunk of the plant, and in it most of its matter and energy are stored; the appendages are organs for taking in food, for excretion, for respiration, for reproduction, etc.

SPECIAL MORPHOLOGY.

Underground Part. The *rhizome* is a hard black, elongated, and branching stem, generally flattened somewhat from above downwards, and expanded slightly on either side to form well-marked lateral folds—the *lateral ridges*. Its width is seldom more than half an inch, and is usually considerably less. In transverse section it has the outline shown in Fig. 35, and the marginal part only is black. The *branches* repeat in all respects the form and structure of the main axis. Both end either in conical, pointed and fleshy structures about two inches long, or in blunt, yellowish knobs, plainly depressed in the centre. At these ends the rhizome grows; hence they are called the growing points or *apical buds* (Figs. 31–34).

Besides the apical buds the rhizome bears nearly always, here and there, a number of dead, decaying tips, which are especially numerous in old plants. They arise in the following manner: After attaining a certain length both the rhizome and its branches die away behind. Death of the hinder part follows at about the same rate with which growth advances at the apical buds; so that the total length does not change materially from year to year. It is obvious that this process must result in the gradual detachment of the branches from the main axis. Each branch, now become an independent rhizome, repeats the process; and in this manner a single original rhizome may give rise to large numbers of distinct plants, all of which have been at some time in material connection with one another. This process is evidently a kind of *reproduction*, though it is not the most important and obvious means for the propagation of the plant. In this way a large area may be occupied by distinct, though related, plants whose branching rhizomes cross and recross, making the subterranean network already described, p. 67.

Arrangement of the Leaves upon the Rhizome. The young plant of *Pteris* puts up a number of leaves (7–12) yearly, but the adult generally develops one only, which grows very slowly, requiring two years before it unfolds. Towards the end of the first year it is recognizable only as a minute knob at the bottom of a depression near the growing point. At the beginning of the second year it is perhaps an inch high, the stalk only having

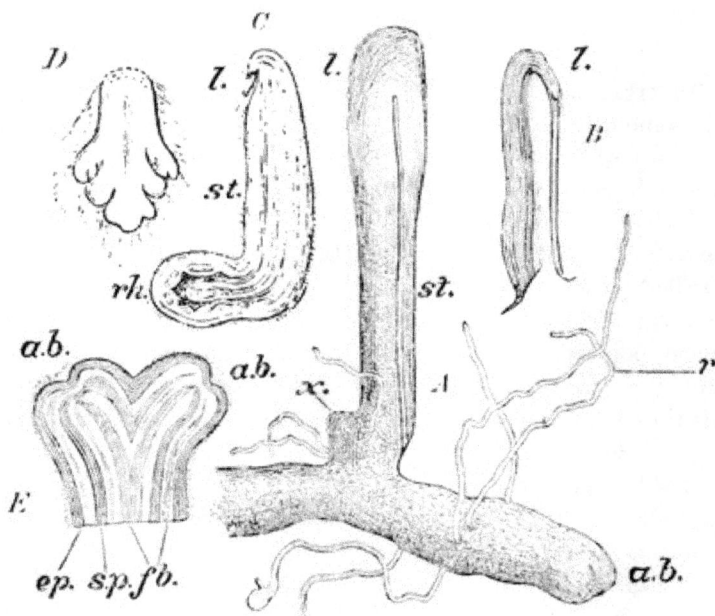

FIG. 33. (After Sachs.)—Developing leaf, etc., of *Pteris*. *A*, end of a branch showing the apical bud and the rudiment of a leaf; *B*, a rudimentary leaf; *C*, a similar leaf in longitudinal section, showing the infolded lamina (*l*), the attachment to the rhizome, and the prolongation of the tissues of the latter into the leaf; *D*, lamina of a very young leaf; *E*, horizontal section through a growing point which has just forked to form two apical buds. *a.b.*, apical bud; *ep*, epidermis and underlying sclerotic parenchyma; *f.b.*, fibrovascular bundles; *l*, lamina; *r*, root; *s.p.*, sclerotic prosenchyma; *x*, an adventitious bud at the base of the leaf.

appeared. At the end of the second year the lamina is developed, and hangs down as shown in Fig. 33, *C.* Early in the spring of the third year it breaks through the ground, and grows rapidly to the fully matured state.

The leaves usually arise near the apical buds of the main

axis or of the branches. Behind each mature leaf remnants of
the leaves of preceding years are often to be found, alternating on
the sides of the rhizome in regular succession, and showing various
stages of decay. The first of these (which is on the opposite side
of the rhizome from the living leaf) was alive the previous year;
the next (on the same side with the living leaf) is the leaf of the
year before that; and so on. Fig. 34 shows an example of this
sort. The leaf of the present year, l^2, is fully developed; and the

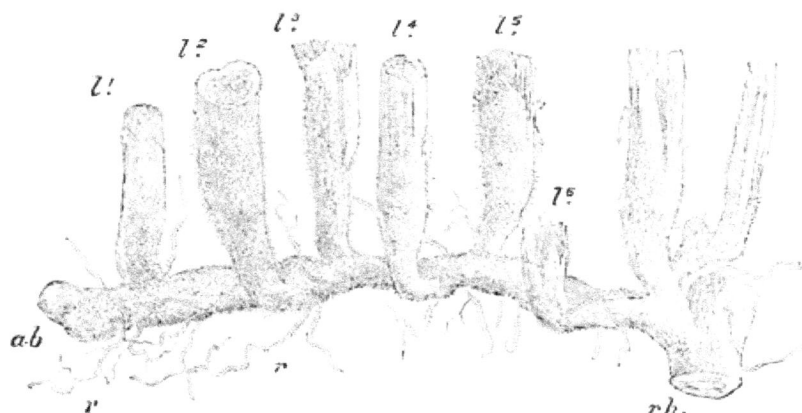

FIG. 34. (After Sachs.)—Branch of a rhizome of *Pteris*, showing the apical bud (*a b*), the stumps of a number of successive leaves (l^1, l^2, l^3, etc.), and a part of the main rhizome (*rh*); *r*, root.

relics of the leaves of the preceding years are indicated at l^2, l^3,
etc.; l^5 is the rudiment of next year's leaf.

Internal Structure of the Rhizome. The rhizome is a nearly
solid mass, consisting of many different kinds of cells, united into
different tissues, and having a very complicated arrangement. Its
study is somewhat difficult. Nevertheless the arrangement of the
cells is definite and constant, and merits careful attention, since
it has many features which are characteristic of the cellular structure of the stems of higher plants. We shall first examine its
more obvious composition as displayed in transverse and longitudinal sections, afterwards making a careful microscopical study
of the cells and tissues.

Seen with a hand-lens or the naked eye, a transverse section of
the rhizome (Fig. 35) presents a white or yellowish background
bounded by a black margin (the *epidermis*) and marked by various

colored or pale spots and bands; the latter are different tissues, or
systems of tissue. These different structures are arranged in
three groups or *systems of tissue*, which are found among all

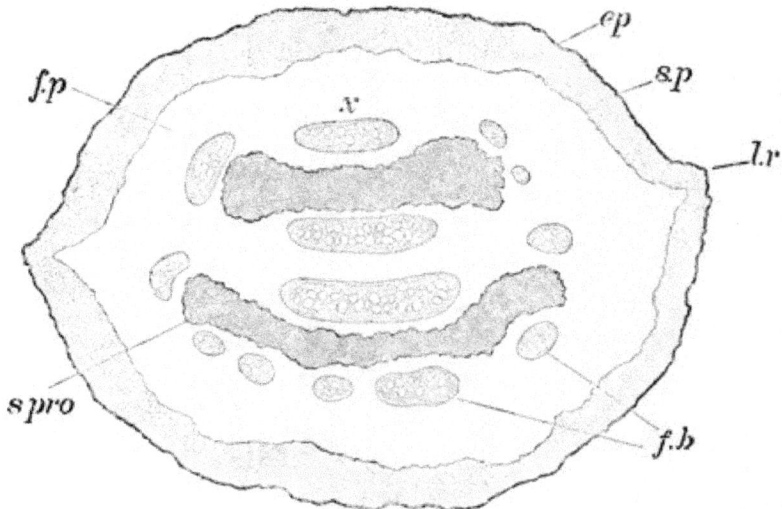

Fig. 35.—Cross-section of the rhizome of *Pteris. l.r*, lateral ridges; *f.p*, fundamental parenchyma; *s.p*, sclerotic parenchyma; *s. pro*, sclerotic prosenchyma; *f.b, x*, fibro-vascular bundles.

higher plants in essentially the same form, though differing
widely in the minor details of their arrangement. These are:—

I. The Fundamental System.

II. The Epidermal System.

III. The Fibro-vascular System.

The *Fundamental system* consists in *Pteris* of three tissues:

(*a*) *fundamental parenchyma* (Fig. 35, *f. p*), the soft whitish
mass forming the principal substance of the rhizome;

(*b*) *sclerotic parenchyma* (*s. p*), the brown hard tissue lying just
below the epidermis, from which it is scarcely distinguishable;

(*c*) *sclerotic prosenchyma* (*s. pro*), black or reddish dots and
bands of extremely hard tissue, most of which is contained in two
conspicuous bands lying one on either side of a plane joining
the lateral ridges.

The sclerotic parenchyma and the sclerotic prosenchyma both

arise through a transformation (hardening, etc.) of portions of originally-soft fundamental parenchyma. In most plants above the ferns the fundamental system contains neither of these tissues.

The *Fibro-vascular system* is composed of longitudinal threads or strands of tissues known as the *fibro-vascular bundles*, and these in one form or another are characteristic of all higher plants. They appear here and there in the section (Fig. 35, *f.b*) as indistinct, pale or silvery areas of a roundish, oval, or elongated shape. Closely examined they have an open texture, enclosing spaces which are sections of empty tubes, or vessels and fibres, from which the bundles take their name.

The *Epidermal system* consists of a single tissue, the *epidermis*, which covers the outside of the rhizome.

By a simple dissection of the stem the sclerotic prosenchyma and the fibro-vascular bundles may be seen to be long strands or bands, coursing through the softer fundamental tissues.

It should be clearly understood that the three systems are in general, not single tissues, but *groups* of tissues which are constantly associated together for the performance of certain functions.*

Minute Anatomy (Histology) of the Rhizome.

General Account. Microscopic study of thin sections of the rhizome shows the various tissues to be composed of innumerable closely-crowded cells, which differ very widely in structure and in function. In studying these cells the student should not lose sight of the fact that they are objects having three dimensions, of which only two are seen in sections. And hence a single section may give an imperfect or entirely false impression of the real form of the cells,—just as the face of a wall of masonry may give only an imperfect idea of the blocks of which it is built. For this reason many of the cells can only be understood by a comparison of

---

* This classification of the tissues is only a matter of convenience, and has little scientific value. By many botanists it has been rejected altogether; but no apology for its use need be made by those who, like the authors, have found it useful, so long as it is defended by Sachs (who first introduced it), and its value for beginners is conceded by De Bary.

transverse and longitudinal sections, and these should be studied together until their relations are thoroughly mastered.

The following table gives brief definitions of the leading vegetal tissues as found in *Pteris* and elsewhere :—

PRINCIPAL ADULT VEGETAL TISSUES.

TISSUES.	CHARACTERISTICS.
1. *Epidermis.*	Cells in a single layer covering the outer surface.
2. *Parenchyma.*	Masses of cells, rounded, prismatic or polyhedral, usually incompletely joined at the angles, thus leaving intercellular spaces. Not much longer than broad. Thin-walled.
3. *Prosenchyma.*	Cells elongated, typically massed, without intercellular spaces.
4. *Sieve-tubes.*	Cells elongated, thin-walled, panelled with perforated areas, containing proteids.
5. *Tracheids.*	Cells thick-walled, elongated, pointed, hard; walls pitted; filled with air.
6. *Vessels.*	Cells very slender, elongated, opening into one another at their ends, often spirally thickened, and filled with air.

These six tissues are not only found in the rhizome, but extend throughout the roots and the fronds as well. Moreover, all the tissues not only of the fern but of all higher plants are varieties of them.

Special Account. It must not be forgotten that the differences between tissues are only the outcome of the differences between their component cells (p. 15). So that the study of the histology of the rhizome, even if preceded (as it may well be) by a dissection, and a naked-eye examination of some of the tissues, eventually resolves itself into the careful microscopic study of the several kinds of cells composing those tissues.

The mature parts of the rhizome contain at least nine very different kinds of cells, the characteristics and grouping of which are shown in the following table. In the apical buds, however, this arrangement disappears, and all the cells become closely similar.

MINUTE ANATOMY OF THE RHIZOME OF *PTERIS AQUILINA.*

SYSTEM.	TISSUES.	CHARACTERISTICS.
I. Epidermal.	1. *Epidermis.*	Cells polygonal in cross-section, empty. Walls hard, thickened, especially towards the outside (Fig. 36.)
II. Fundamental.	2. *Fundamental parenchyma.*	Cells rounded or polygonal in cross-section, colorless. Thin-walled, containing protoplasm, nucleus and starch. Intercellular spaces present. (Fig. 39, *f.p.*)
	3. *Sclerotic parenchyma.*	Cells polygonal or semi-fusiform in section, nearly empty. No intercellular spaces. Walls hard and brown, thickened. (Fig. 36.)
	4. *Sclerotic prosenchyma (or sclerenchyma).*	Cells fusiform, empty. Walls thick, red. (Fig. 37.)
III. Fibro-vascular.	5. *Wood-parenchyma.*	Like the fundamental parenchyma, but with more elongated cells. (Figs. 39, 40.)
	6. *Phloëm-parenchyma.*	Precisely like 5, differing only in position.
	7. *Phloëm prosenchyma, or bast-fibres.*	Cells fusiform, rich in protoplasm, colorless. Walls thick, soft. (Figs. 39, 40.)
	8. *Sieve-tubes.*	Having the ordinary characters (see preceding table). (Figs. 39-41.)
	9. *Tracheïds (ladder-cells).*	Pits transversely elongated (scalariform). (Figs. 38, 39, 40.)
	10. *Vessels (spiral).*	Very slender, with one or two internal spiral thickenings. (Fig. 39.)

Besides the above-mentioned tissues, the rhizome contains certain other secondary varieties which will be described further on.

Epidermal System. Epidermis. It is the function of the epidermis (aided in this case by the underlying sclerotic parenchyma) to protect the inner tissues from contact with the soil and to guard against desiccation of the rhizome during droughts. The cells (Fig. 36) are dead and empty, with enormously thick, hard walls perforated by numerous branching canals. The outer wall is especially thick.

Fundamental System. The tissues of this system form the main body of the plant, and in the fern have two widely different functions. The *fundamental parenchyma* is a kind of store-

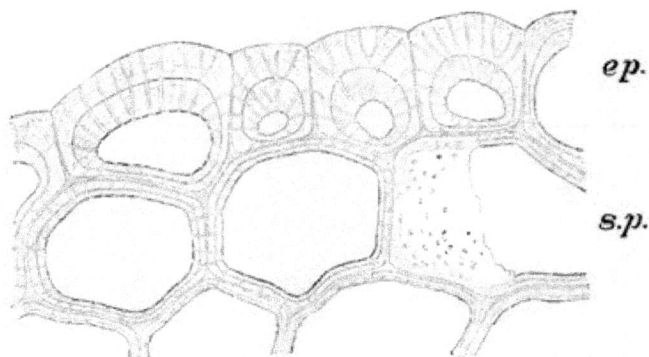

FIG. 36.—Section showing the epidermis (*ep*), and the underlying sclerotic parenchyma (*s.p*) of the rhizome of *Pteris aquilina*. Canals, sometimes branching, are everywhere seen.

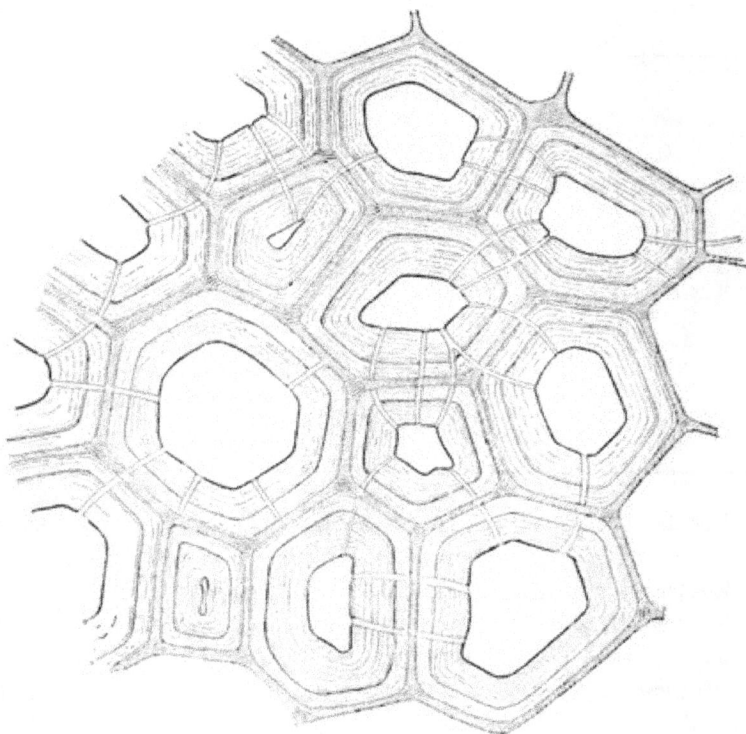

FIG. 37.—Cross section of sclerotic prosenchyma of the rhizome of *Pteris aquilina*. The enormously thickened walls consist of three layers, are perforated by pores, and are *lignified* or turned into wood.

house in which matter and energy are stored—mainly in the form of starch, $C_6H_{10}O_5$,—and in which active chemical changes take place. The cells are thin-walled and soft, and are rather loosely joined together, leaving numerous intercellular spaces (Figs. 39, 40). They contain protoplasm and nucleus, and very numerous rounded grains of starch. This starch is stored up by the plant during the summer as a reserve supply of food—just as hibernating animals store up fat in their bodies for use during the winter. Accordingly, it increases in quantity during the summer and decreases in the spring when the plant resumes its growth, before the leaves are unfolded. The parenchyma has also the function of conducting various substances (especially dissolved sugar) through the plant by diffusion from cell to cell.

The *sclerotic parenchyma* and *sclerotic prosenchyma* (Figs. 36, 37) are dead, and hence play a passive part in the adult vegetal economy. The former co-operates with the epidermis; the latter probably serves in part to support the soft tissues, and to some extent affords a channel for the conveyance of the sap. The sap, however, does not flow through the cavities, but passes slowly along the substance of the porous walls. The cells of both these sclerotic tissues have very thick, hard, brown walls, perforated here and there by narrow channels. The cells of the parenchyma are prismatic or polyhedral; those of the prosenchyma elongated, and pointed at their ends. In both, the protoplasm and nuclei disappear when the cells are fully formed. Towards the apical buds both fade into ordinary fundamental parenchyma.

Fibro-vascular System. The *fibrovascular bundles* (p. 73) are long strands or bands of tissue which appear in cross-section as isolated

Fig. 38. (After Sachs.)—View of the rhizome, which is supposed to be transparent so as to show the network of the upper fibrovascular bundles. *l*, a leaf.

spots (Fig. 35). The bundles are not really isolated (p.   ), however, but join one another here and there, forming an

open network (Fig. 38), which can only be seen in a lateral
view of the rhizome. From this network bundles are given off
which extend on the one hand into the roots and on the other
into the leaves, branching in the latter to form the complicated
system of veins to be described hereafter (p. 87).

Each bundle consists of a number of different tissues which,
broadly speaking, have the function of conducting sap from one
part of the plant to another. These tissues have the following

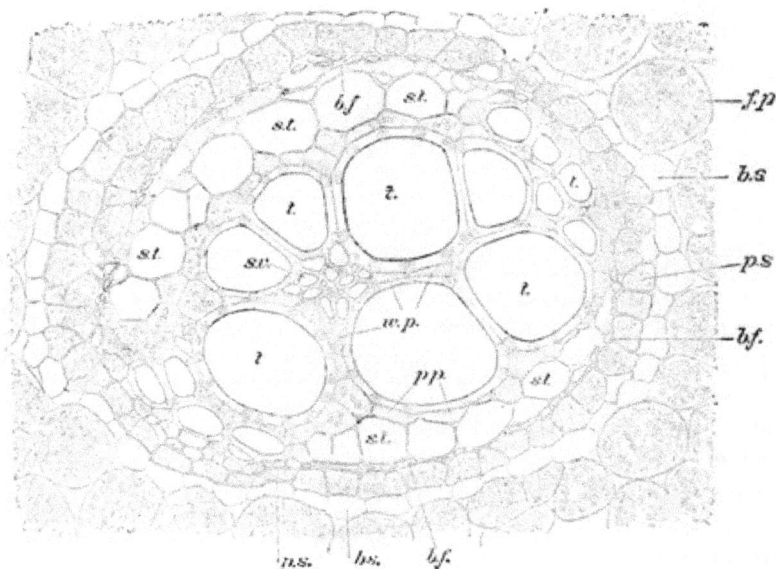

FIG. 39.— Highly magnified cross-section of a fibro-vascular bundle surrounded by the fundamental parenchyma. *f.p. t.* scalariform tracheids; *b.s.* bundle-sheath; *p.s.* phloëm-sheath; *b.f.* bast-fibres; *s.t.* sieve-tubes; *p.p.* phloëm-parenchyma; *w.p.* wood (xylem) parenchyma; *s.v.* spiral vessel.

definite arrangement. Beginning with the outside of a bundle,
we find (Figs. 39, 40)—

1. *Bundle sheath ;* a single layer of elongated cells enveloping the bundle, probably derived from and belonging to the
fundamental system.

2. *Phloëm-sheath ;* a single layer of larger parenchymatous
cells containing starch in large quantities.

3. *Bast-fibres ;* soft, thick-walled, elongated, pointed cells containing protoplasm and large nuclei.

4. *Sieve-tubes ;* larger, soft, thin-walled, elongated cells con-

taining protoplasm and having the walls marked by areas perforated by numerous fine pores (panelled). They join at the ends by oblique panelled partitions (shown in Figs. 40 and 41).

5. *Phloëm-parenchyma;* ordinary parenchymatous cells filled with starch, scattered here and there among the bast-fibres and sieve-tubes.

6. *Tracheids (scalariform)* or "ladder-cells;" occupying most of the central part of the bundle. Their structure calls for some remark. They are empty or air-filled fusiform tubes, whose hard, thick walls are in the young tissue sculptured with great regularity into a series of transverse hollows or pits, which finally become actual holes. The walls of the tracheid are therefore

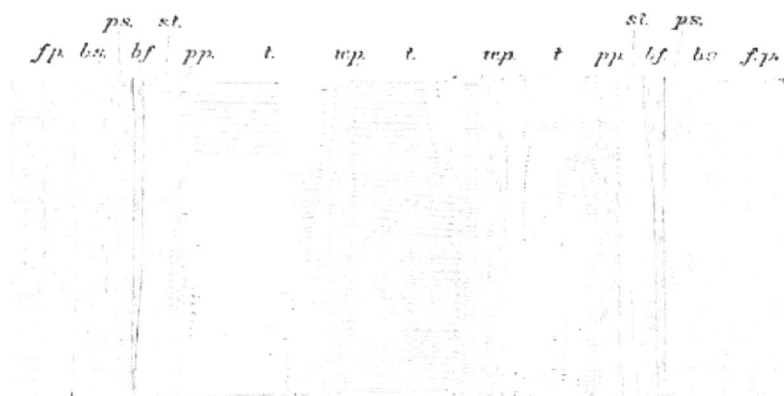

FIG. 40.—Longitudinal section of a fibro-vascular bundle, surrounded by the fundamental parenchyma. *b.f.* bast-fibres; *b.s.* bundle-sheath; *f.p.* fundamental parenchyma; *p.p.* phloëm-parenchyma; *p.s.* phloëm-sheath; *s.t.* sieve-tubes; *t.* scalariform tracheids or ladder-cells; *w.p.* wood-parenchyma.

continuous at the angles, but along their plane surfaces become converted into a series of parallel bars, making a grating of singular beauty. The slits between the bars are not rectangular passages through the wall, but are rather like elongated, flattened funnels, opening outwards. The sides of the funnels are called the *borders* of the *pits;* and pits of this sort are called *bordered scalariform pits* (cf. Figs. 29 and 40).

7. *Vessels (spiral);* scattered here and there among the tracheids, and hardly distinguishable from them in cross-section. They are continuous elongated tubes filled with air, and strengthened by a beautiful close spiral ridge (sometimes two) which runs round the inner face of the wall (Fig. 39).

The tracheids and vessels are of great physiological importance, being probably the main channels for the flow of sap.
Sap is water holding various substances in solution. The water
enters by the roots, *flows principally through the walls of the
vessels and tracheids, and not through their cavities, which are
filled with air*, and is thus conducted through the rhizome and
upwards into the leaves.

8. *Wood-parenchyma ;* cells like those of the phloëm-parenchyma (5) scattered between the vessels and tracheids.

Branches of the Rhizome. These repeat in all respects the
structure of the main stem. They are equivalent members of
the underground part, and differ in no wise, excepting in their
origin, from the main stem itself.

Roots. The roots may easily be recognized by their small
size and tapering form, and their lack of the lateral ridges of the

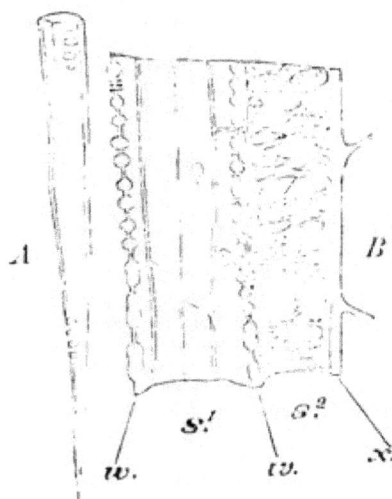

Fig. 41. (After De Bary.)— Sieve-tubes from the rhizome of *Pteris aquilina*, showing : *A*, the
end of a member of a sieve-tube; *B*, part of a thin longitudinal section. The section has approximately halved two sieve-tubes, S_1 and S_2, which are so drawn that the
uninjured side lies behind. The broad posterior surface of S_2 is seen covered with
sieve-plates connecting with another sieve-tube. S_1, on the contrary, abuts by a smooth
non-plated surface upon parenchymatous cells which are seen through it. *w*, sections
of walls bearing sieve-pits; *x*. section of a non-plated wall abutting upon parenchyma.

stem and branches. They arise *endogenously* from the main
stem or its branches, i. e., by an outgrowth of the internal tissues

and not (as in the case of the false roots or *rhizoids* of the prothallium, shortly to be described) by elongation of superficial
cells of the epidermis. True roots, of which those of *Pteris* are
good examples, arise always as well from the fundamental and
fibro-vascular regions, and include all the systems found in the
stem itself. Hence cross-sections of *Pteris* roots differ but slightly
from those of the stem or the branches, and the root in general
is clearly a member of the plant body. As in all true roots, the
free end is covered by a special boring tip called the *root-cap*, but
this is apt to be lost in removing the specimen from the earth.

Growth of the Rhizome.

The Embryonic Tissue or Meristem. The starting-point in the
development of the underground parts will be fully described
hereafter (p. 98). But briefly stated it is as follows: By continued division of the germ-cell (compare p. 51) a mass of similar
cells is formed, part of which is destined to develop into the rhizome. By gradual differentiation (p. 15) of the cells, this portion
is converted into the mature rhizome, which however remains
nearly undifferentiated at the tip. At this point the epidermis
may be distinguished, but it remains very delicate, and the underlying cells continue to grow and multiply, thus producing continued elongation of the mass. In this way the first apical bud
is formed. Lateral buds are then given off right and left to constitute the embryos of leaves, branches or roots, which, still re-
taining their soft and delicate tips, are capable of further growth.

Behind these "growing points" the epidermis and other
tissues grow more and more slowly, and soon reach their maximum size, whereupon rapid growth ceases. The power of growth
is henceforward mainly confined to the apical buds, and the growing tissue of which they are composed is known as *embryonic
tissue* or *meristem*.

The Apical Cell. Close examination reveals the fact that each
apical bud contains a remarkable cell which is especially concerned
in the function of growth, viz., the *apical cell*, which lies in a hollow at the apex of the bud. In the apical buds of the rhizome or
branches this cell has somewhat the form of a wedge-shaped
pyramid. It continually increases in size, but as it grows, repeat-
edly divides so as to cut off cells behind it alternately on its right

6

and left sides. These cells in turn continue to grow and divide and thus give rise to two similar masses of meristem, which together constitute the apical bud. From the meristem by gradual, though rapid, changes the various tissues of the adult rhizome are differentiated; and longitudinal sections passing through

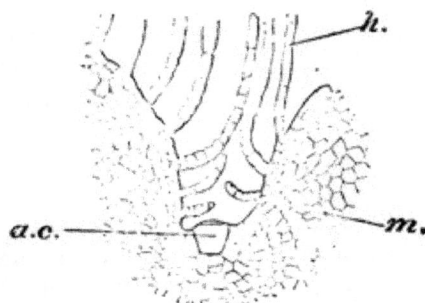

Fig. 42. (After Hofmeister.)—Apical cell of the rhizome in a vertical longitudinal section. *a.c,* apical cell; *h,* hair; *m,* meristem.

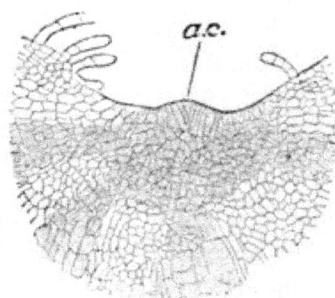

Fig. 43. (After Hofmeister.)—Apical cell of the rhizome in horizontal longitudinal section. *a.c,* apical cell.

the lateral ridges show the mature tissues fading out in a region of indifferent meristem about the apical cell (Fig. 43).

The apical cell lies at the bottom of a funnel-shaped depression at the tip of the stem. It is shaped approximately like a thin, two-edged wedge with an arched or curved base turned forwards towards the centre of the funnel-shaped depression. The thin edge of the wedge is directed backwards, and its sides, which are also curved, meet in a vertical plane above and below. A longitudinal section taken through the plane of the

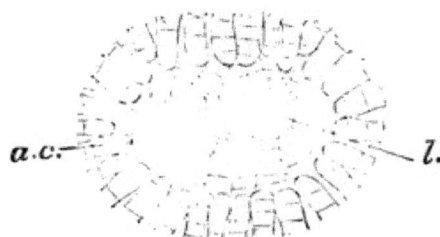

Fig. 44. (After Sachs.)—A vertical transverse section through the *apical cell, a.c,* showing a boundary of hairs and a second apical cell, *l,* belonging to a leaf.

lateral ridges therefore shows the apical cell in a triangular form as in Fig. 43. A section taken at right angles to this—*i.e.,* vertical and longitudinal,—shows the cell to be approximately rectangular and quadrilateral (Fig. 42), while a transverse vertical section shows it in the form of a bi-convex lens (Fig. 44).

The funnel-shaped depression is compressed vertically, and its walls are thickly covered with erect branching hairs which are closely fastened together by a hardened mucilage secreted by the apical bud. These hairs entirely close the mouth of the funnel and shut off the delicate young

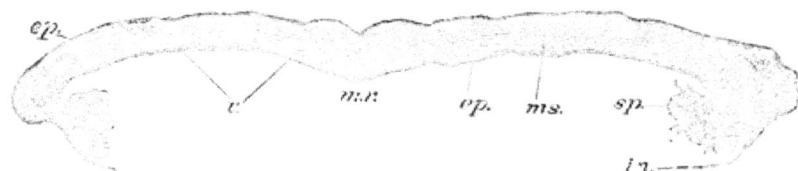

FIG. 45.—Cross-section of an entire fertile leaflet. *m.r*, midrib; *r*, veins; *ep*, epidermis; *ms*, mesophyll; *sp*, sporangia; *in*, indusium.

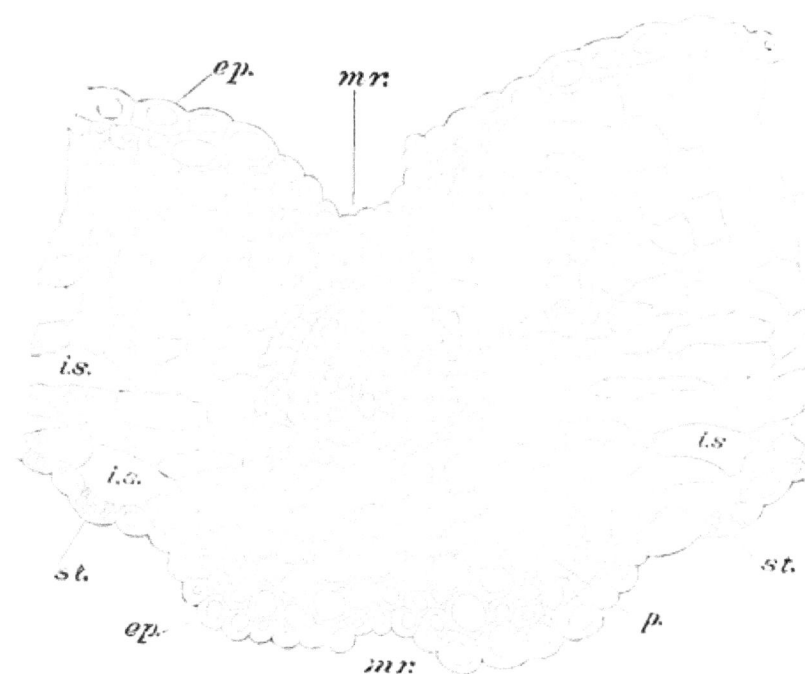

FIG. 46.—Cross-section, still more enlarged, passing through the midrib of a leaflet. In the centre the circular fibro-vascular bundle, supported, especially above and below, by thickened prosenchyma (*p*). On either side the parenchymatous mesophyll cells (shaded) and the intercellular spaces (*i.s*) opening by stomata (*st*); epidermis *ep*.

portions at its base from the outer world. Protected by these hairs, the end of the stem forces its way through the toughest clay without injury to the delicate bud buried in its apex. (Hofmeister.)

Aerial Part of the Brake. Leaf.

The external form of the leaf has been described on p. 68, and it now remains to consider its internal structure. The lamina is to be regarded as an expanded portion of the stipe, made thin and delicate in order to present a large surface to the light and the air. The stipe, in turn, is a prolongation from the rhizome, so that the whole plant is a continuous mass, throughout which extend the three systems of tissue virtually

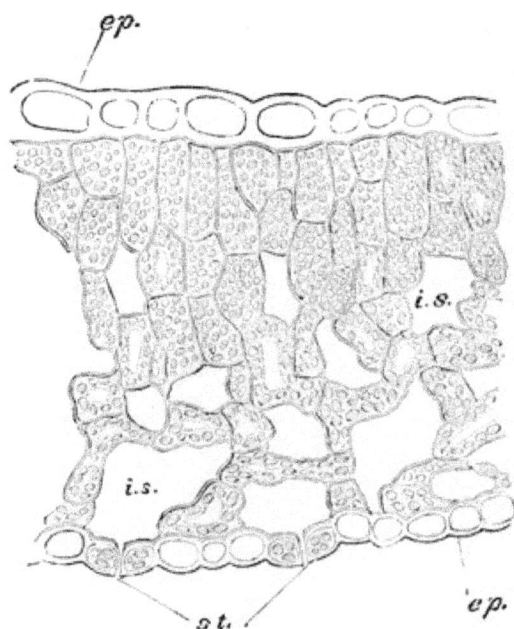

Fig. 47.—Cross-section of part of a leaflet showing the microscopic structure. *ep*, epidermis; *st*, stomata; *i.s.* intercellular spaces between the mesophyll-cells, which are filled with (shaded) chlorophyll-bodies lying in the protoplasm.

unchanged. The transverse and longitudinal sections of the stipe show only minor points of difference from corresponding sections of the rhizome. In the leaf, however, all three systems undergo great changes. The epidermis becomes very thin, delicate, and transparent; the fibro-vascular bundles break up into an extremely fine and complex network forming the *veins*; the sclerotic tissues become transparent and are found only along the veins. The cells of the fundamental parenchyma alter their form, lose their starch, and become filled with bright-green, rounded bodies, called the *chlorophyll-bodies*, which

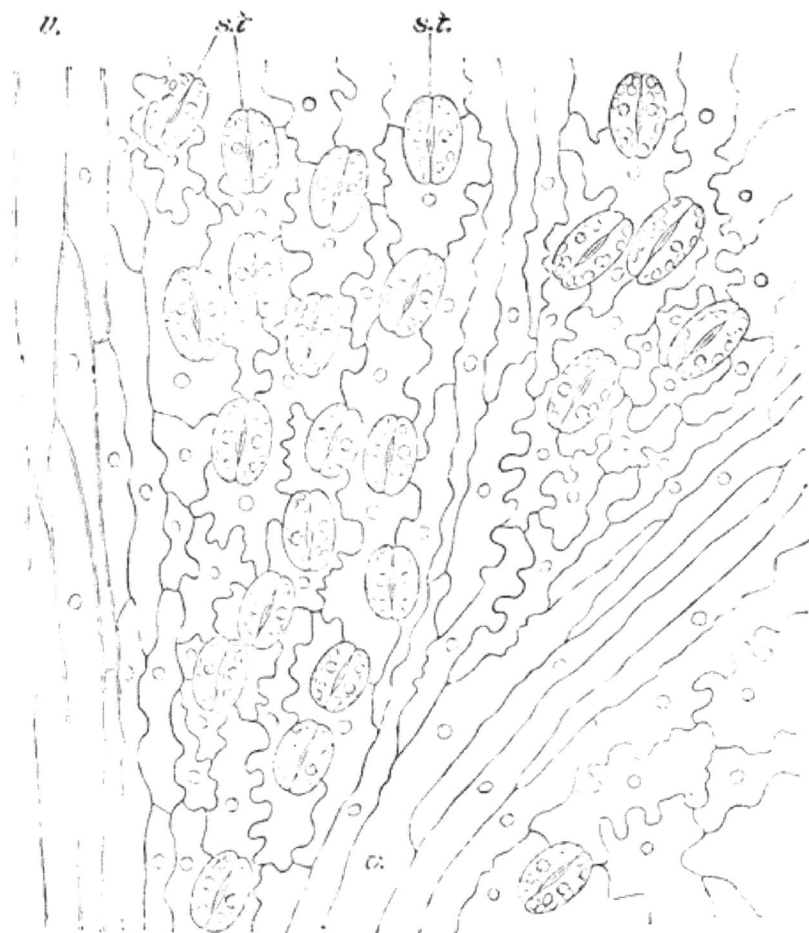

FIG. 48.—Epidermis from the under side of a leaflet, showing wavy cells; elongated (*prosenchymatous*) cells over the veins; and stomata with their guard-cells. *st*, stomata and guard-cells; *v*, veins covered by thick and prosenchymatous epidermal cells. Intermediate stages between wavy and straight cells are also shown. (Surface view.)

are composed of a protoplasmic basis colored by a substance known as **chlorophyll**. The green parenchyma is often called the *mesophyll*.

A cross-section of a leaflet (p. 68) is shown in Fig. 45. The finer structure of the leaflet is shown in Figs. 46 and 47. On the outside is the epidermis (*ep*); within, the mesophyll and mid-rib—the latter composed of thickened epidermal and sclerotic fundamental tissue, and a large fibro-vascular bundle.

The mesophyll, or leaf-parenchyma, consists of irregular cells which are loosely arranged on the lower side leaving very large intercellular spaces, but are closely packed, and leave few or no intercellular spaces, on the upper (sunny) side. The cells have very thin walls, contain protoplasm and a large central space (vacuole) filled with sap, and numerous chlorophyll-bodies imbedded in the protoplasm. These are especially numerous in the upper part of the leaf, as might be expected from their functions in connection with the action of light (see page 106).

The epidermis, or *skin* of the leaf, consists of translucent, greatly flattened cells having peculiar wavy outlines and rela-

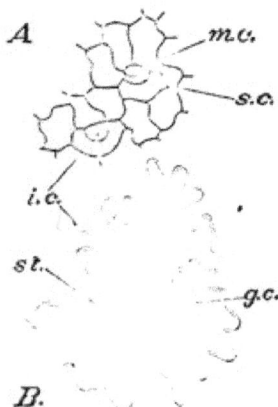

Fig. 49. (After Sachs.)—Epidermal cells of *Pteris flabellata*, showing the development of stomata. *A.* very young epidermal cells: *B.* nearly mature: *i.c.* initial cell: *m.c.* mother-cell: *s.c.* subsidiary cell: *g.c.* guard-cell: *st.* stoma.

tively thick walls (Figs. 36, 46–49). Upon the veins they become elongated, and their walls are considerably thickened, especially upon the midrib (Figs. 46, 48). They generally contain large, distinct nuclei, and often considerable protoplasm. The wavy epidermal cells, particularly in young plants, contain some chlorophyll and starch, though in this respect the fern is somewhat exceptional.

In the rhizome the epidermis forms a continuous layer over the whole surface. In the leaf, however, this is not the case, the epidermis on the lower side being perforated by holes leading into the interior and known as mouths or *stomata* (singular, *stoma*) (Fig. 48). These holes do not pass into the cells, but are gaps or breaks between certain cells of the epidermis, and open directly into the intercellular spaces of which they are, in fact, the ends. That portion of the intercellular labyrinth which directly underlies the stoma is sometimes called the respiratory cavity. Each stoma is bounded, as in most plants, by two curving *guard-cells*, which are generally nucleated, and, unlike epidermal cells generally, contain abundant chlorophyll-bodies and starch.

The guard-cells are capable of changing their form according to the amount of light, the hygroscopic state of the atmosphere, and other circumstances, and thus open or close the hole or stoma between them. This action is of great importance in the physiology of the plant (transpiration, p. 105).

In *Pteris cretica* and *flabellata* the stomata develop as follows : A young epidermal cell is divided by a curved partition into two cells, one of which (Fig. 49) is called the *initial cell* of the stoma (*i.c*). This is again divided by a curved partition into the mother-cell of the stoma (Fig. 49, *m.c*) and a subsidiary cell (Fig. 49, *s.c*).

The mother-cell is then bisected into the two *guard-cells*, and the stoma appears as a chink between them (Fig. 49, *B*).

The *veins* are the fibres or threads which constitute the framework of the leaf. Each consists, essentially, of a small fibrovascular bundle branching from that of the midrib (Figs. 45, 46, 50). Above and below them, the mesophyll and epidermal cells are generally thickened and prosenchymatous, in this way contributing alike to the form and the function of the " vein."

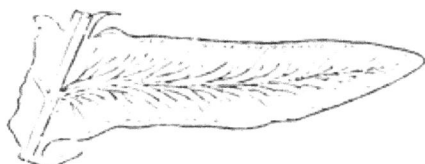

Fig. 50. (After Luerssen.)—Venation of a leaflet of *Pteris aquilina.*

Their arrangement (veining or *venation*) is definite, and depends on the mode of branching of the fibro-vascular strand which constitutes the principal part of the midrib. Secondary strands (nerves) proceed from this at an acute angle, then turn somewhat abruptly towards the edge of the leaflet (or lobe), making an arch which is convex towards the distal extremity of midrib (Fig. 50).

From this point, after branching once or twice, the delicate veins run parallel to each other to the edge of the leaflet, where they join one another or *anastomose.* This form of venation is known as *Nervatio Neuropteridis*, and is more easily seen in the leaf of *Osmunda regalis* (cf. Luerssen. *Rabenhorst's Kryptogamen-Flora* (1884). III., s. 12).

CHAPTER VI.

THE BIOLOGY OF A PLANT (*Continued*).

THE BRAKE OR FERN.

Reproduction. The highly important *generative* function is performed in *Pteris* by structures which are very unlike and somewhat complicated. The subsequent descriptions of these organs will be better understood if prefaced by a brief account of reproduction in general, as well as in the special case of the fern.

Living things give rise to offspring by the detachment of certain portions or fragments of their own bodies, endowed with the power to grow and develop into complete individuals; and so far as known, living things never arise except through this process. The detached portions are sometimes masses of cells (*vegetative* reproduction), as in the separation of whole branches or of special, so-called *adventitious*, buds, but more commonly are single cells (like the *ova* of animals and the *spores* of ferns and mosses).

All forms of reproduction fall under one or the other of two heads, viz., **Agamogenesis** (*asexual* reproduction) or **Gamogenesis** (*sexual* reproduction). In the former case the detached portions have the power to develop into new individuals without the influence of other living matter. In the latter, the detached portion, in this case always a single cell called the *germ* (ovum, oösphere, etc.), is acted upon by a different portion of living matter, likewise a single cell, which in most cases has been detached from the body of another organism. The germ is called the *female germ-cell*; the cell acting upon it the *male germ-cell*; and in the sexual process the two fuse together (*fertilization, impregnation*) to form a single cell endowed with the power of developing into a new individual. In some organisms (e.g., the yeast-plant) only agamogenesis has been observed; in others (e.g., vertebrates) only gamogenesis. In still others, both processes take place, and of such organisms the fern is a conspicuous example.

Pteris possesses two modes of "vegetative" asexual reproduction, viz., the separation of entire branches from the rhizome, as already mentioned (p. 69), and of "adventitious buds" from the bases of the leaf-stalks (Fig. 33). But besides these, the fern has a quite different method of agamogenesis which regularly alternates with a process of gamogenesis (*alternation of generations*). The following brief outline of this important process may help to guide the student through the subsequent detailed descriptions.

Upon some of the leaves are formed organs called *sporangia* (Figs. 45, 51, 52), which produce numerous reproductive cells called *spores*. The spore becomes separated from the parent plant and develops into an independent plant, the *prothallium* (Fig. 58), which differs entirely in appearance from the fern, and ultimately produces male and female germ-cells. The female cell is fertilized by a male cell, and thereupon develops into an ordinary "fern," which in turn produces spores. The formation and development of the spores is evidently a process of *agamogenesis*, and the fern proper is therefore neither male nor female—i.e., it is *asexual*.

The formation and development of the germ-cells, on the contrary, is a process of *gamogenesis*; and the prothallium is a distinct sexual plant, being both male and female (*hermaphrodite* or *bisexual*). In general terms this is expressed by calling the asexual fern the spore-bearer, or *sporophore*, and the prothallium the egg-bearer, or *oöphore*. The life-history of the fern, broadly speaking, consists therefore in the regular alternation of the *sporophore* (asexual generation) with the *oöphore* (sexual generation), that is, it consists of an *alternation of generations*. An essentially similar alternation of sporophore with oöphore occurs in all higher plants, though in most cases it is so disguised as to escape ordinary observation.

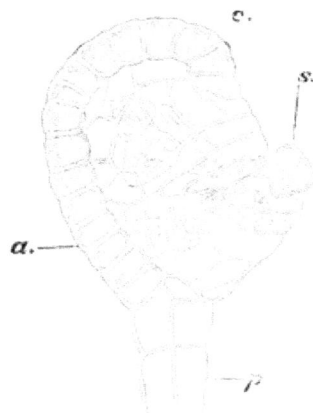

Fig. 51. (After Sminski.) Sporangium of *Pteris serrulata*. *p*, pedicel; *c*, capsule; *a*, annulus; *s*, spore.

The Sporangia and Spores. The *sporangia* of ferns (Fig. 51)

were discovered in 1648 by Cæsius. They arise in *Pteris* upon a
longitudinal thickening of tissue situated on the under side of the
leaflets near their edges, and including a marginal anastomosis
of the veins. This swelling is known as the *receptacle.* Hairs
are not uncommon upon the under side of the leaf, and some
are found upon or near the receptacle. On the latter arise structures at first superficially similar to hairs, which become enlarged

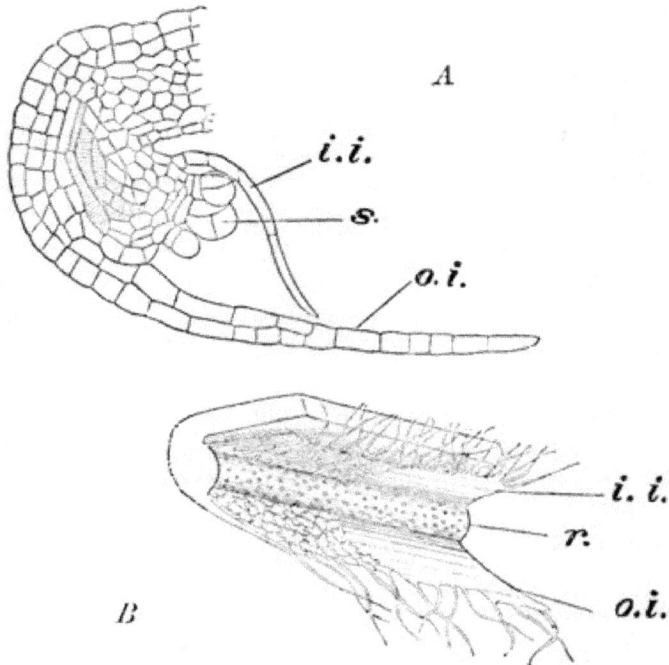

FIG. 52. (From Laerssen, after Burck.)—Indusia and receptacle of *Pteris aquilina;* B
(diagrammatic, seen from below; *A,* in the section of the edge of a leaflet. *o.i,* outer
(false) indusium; *i.i,* inner (true) indusium; *r.* receptacle; *s,* young sporangia.

at the tip, and finally develop into the sporangia. Meanwhile
the edge of the leaflet is bent down and under so as to make
a longitudinal band of thin tissue composed of epidermis known
as the *outer veil* or *indusium* (Fig. 52, *B, o.i*). A similar
thin sheet of epidermis grows down from the under side of
the leaf, and passing outwards to meet the former, constitutes
the *inner veil* or *true indusium* (Fig. 52, *B, i.i*).

In the V-shaped space thus formed the sporangia are developed

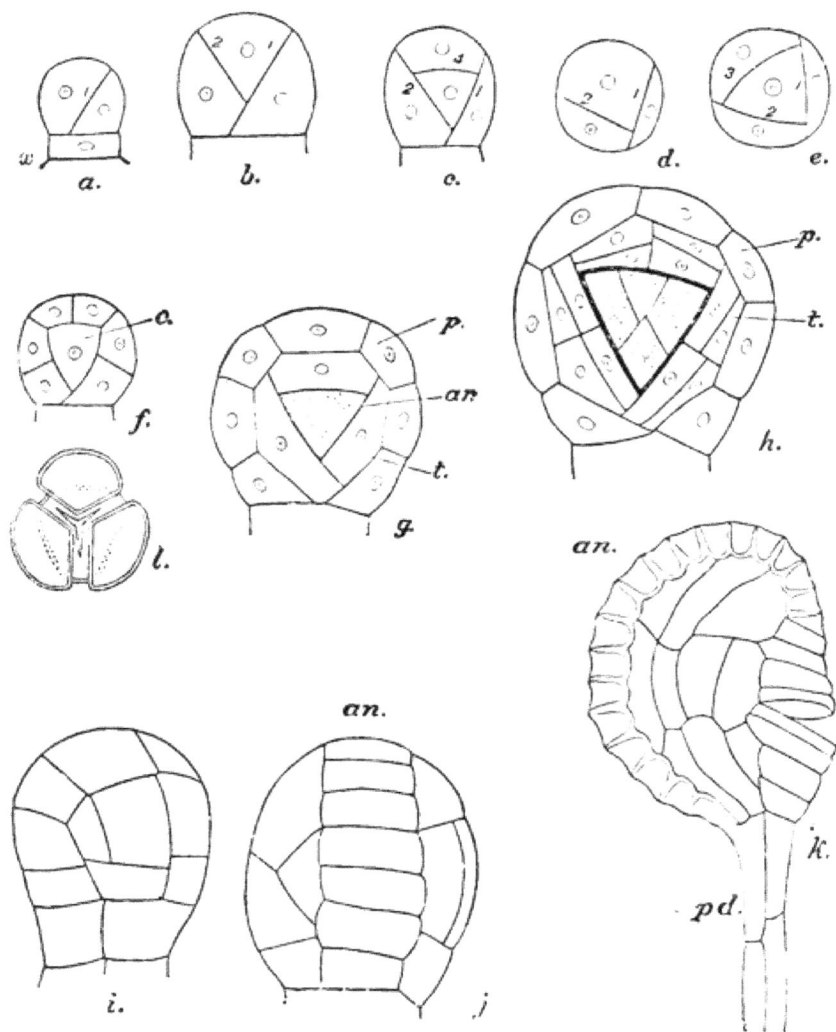

FIG. 53. (After Luerssen.)—Development of the sporangia of *Aspidium Filix mas*, which is closely similar to that of *Pteris*. *a*, the young sporangium standing upon the epidermis-cell from which it has just been divided; *x*, the proximal cell cut off from the sporangium to form the pedicel and support the capsule. *a*, 1, the first partition in the capsule; *b*, 1 and 2, the first and second partitions; *c*, 1, 2, 3, the first, second, and fourth partitions; *d* and *e* are cross-sections of the capsule showing the oblique position of the partitions, and especially that of the third; *f*, a later stage; *g*, the origin of the tapetal cells and the formation of the archesporium; *h*, division of the tapetal cells and the formation of the spore mother-cells; *l*, four spores as they originate in the spore mother-cells; *i*, *j*, *k*, the annulus and ripe sporangium, in surface view. *p*, peripheral cells; *ar*, archesporium; *t*, tapetal cells; *an*, annulus.

and matured in the following manner : A superficial (epidermal) cell enlarges and becomes divided into a proximal (basal) cell and a distal (apical) cell (Fig. 53, *a*). The former develops into the future *pedicel* or stalk of the sporangium ; the latter gives rise to the head or *capsule* within which the spores are formed (cf. Fig. 51). The pedicel arises from the original pedicel-cell by continued growth and subdivision until it consists of three rows of cells somewhat elongated. The rounded capsule-cell is next transformed by four successive oblique divisions into four plano-convex " parietal cells" and a tetrahedral central cell, the *archesporium*, enclosed by the others. The capsule-cell is thus divided by three planes inclined at about 120° (Fig. 53, *b*, *c*). A fourth (Fig. 53, *d*, *e*) passes nearly parallel to the top of the capsule and cuts off from it the central cell or archesporium. In the parietal cells further divisions follow, perpendicular to the surface, while the archesporium gives rise to four intermediate or *tapetal* cells, parallel to the original parietal group (Fig. 53, *g*). The sporangium now consists of a central tetrahedral archesporium bounded by four tapetal cells, which in turn are enclosed by the parietal cells, at this time rapidly multiplying by divisions perpendicular to the exterior. Owing to the peculiar position of the planes of division the whole capsule is now somewhat flattened, and it becomes still more so by the formation along the edge, of a peculiar structure called the *ring* or *annulus*, whose function is the rupturing of the capsule and the liberation of the spores. The annulus is formed by a number of parallel transverse partitions (Fig. 53, *f*, *h*, *i*, *j*), which subdivide the peripheral cells of one edge of the capsule until a certain number of cells have been formed. These then project upon the capsule (Fig. 53, *j*) and form an incomplete ring (Fig. 53, *k*).

Meanwhile the tapetal cells sometimes subdivide so as to form a double row (Fig. 53, *k*, and soon afterwards are absorbed, space being thus left for the growth and enlargement of the archesporium. The latter now divides—first into 2, then into 4, 8, and finally 16 cells, the *mother-cells of the spores*. These remain for a time closely united, but eventually separate and again subdivide, each into 4 daughter-cells (Fig. 53, *l*). The 64 cells thus formed are the asexual *spores*. In their mature state they have a tetrahedral form and certain external markings indicated in Figs. 51, 54. Each spore acquires a double membrane, viz., an inner, *endosporium*, delicate and white, and an outer, "*exosporium*," yellowish-brown, hard, and sculptured over the surface with very close and fine, but irregular, warty excrescences.

Germination of the Spores. Origin of the Prothallium. In the brake the spores ripen in July or August and are set free by rupture of the sporangium under the strain exerted by the elastic annulus, as indicated in Fig. 51. Germination of the spores normally occurs

FIG. 54. (After Suminski.)—Single spore of *Pteris serrulata*.

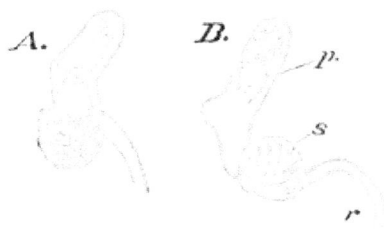

FIG. 55. (After Suminski.)—Germinating spores of *Pteris serrulata*. *A*, in an early stage; *B*, after the appearance of one transverse partition; *s*, spore; *p*, protonema; *r*, rhizoid.

FIG. 56. (After Suminski.)—Very young prothallium of *Pteris*, showing the spore *s*, two rhizoids *r*, and the enlarging extremity.

only after a considerable period (perhaps not before the following spring); it begins by a rupture of the exosporium which is probably immediately due to an imbibition of water. The spore bursts irregularly along the borders of the pyramidal surfaces, and from the opening thus formed the endosporium protrudes as a papilla filled with protoplasm in which numerous chlorophyll-bodies soon appear.

This papilla is known as the *protonema*, or first portion of the prothallium. It develops very quickly (Fig. 55) into a stout cylindrical protrusion divided into cells joined end to end. Close to the spore one or more *rhizoids* are put down from the growing protonema to serve as anchors and roots. At the opposite or distal end longitudinal partitions soon appear (Fig. 56), which speedily convert this portion into a broad flat plate at first only one cell thick, but eventually several cells thick along the median line. This thickening is the so-called "cushion" (see Fig. 58). The whole prothallium is now somewhat spatulate (Fig. 57), but by further growth in front, by an apical cell or otherwise, the wider end becomes still more flattened and heart-shaped or even kidney-shaped. Numerous rhizoids (so-called because they are not morphologically true roots) are put down, and the whole structure assumes approximately the appearance indicated in Fig. 58. The spore-membranes and protonema soon fall away and the prothallium enters upon an independent existence, being rooted by its rhizoids and having an abundance of chlorophyll. In the broad thin plate of tissue no subdivision into stem and leaf exists, and the plant-body closely resembles the "thallus" of one of the lowest plants. Since it is the precursor of the ordinary "fern," it is called the "*prothallus*" or "*prothallium*." *

The cushion forms a prominence on the lower side; upon its posterior part most of the rhizoids are borne.

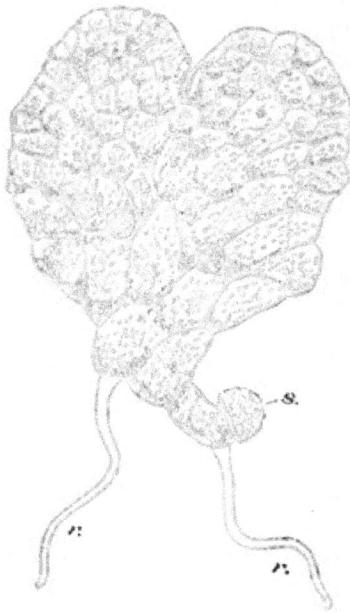

FIG. 57. (After Suminski.)—Older prothallium, showing two rhizoids, three young antheridia and numerous chlorophyll-bodies.

Sexual Organs of the Prothallium.

The prothallia of ferns are as a rule bisexual or double-sexed; that is, each individual possesses both male and female organs. But the latter appear somewhat later than the former, and poorly nourished prothallia often bear only male organs, though they will frequently develop female organs also if placed in better circumstances.

The *Antheridia*, or male organs, first recognized in ferns by Nägeli in 1844, are hemispherical prominences occurring upon the posterior part and the under side of the prothallium, often among the rhizoids. When fully formed (Fig. 58, Fig. 59) an antheridium consists of a mass of rounded cells (*spermatozoid mother-cells*) enveloped by a membrane one cell in thickness.

* The prothallium of ferns was discovered by Ehrhart in 1788.

The mode of origin of the mother-cells differs considerably in different ferns, but in all cases is essentially as follows: An ordi-

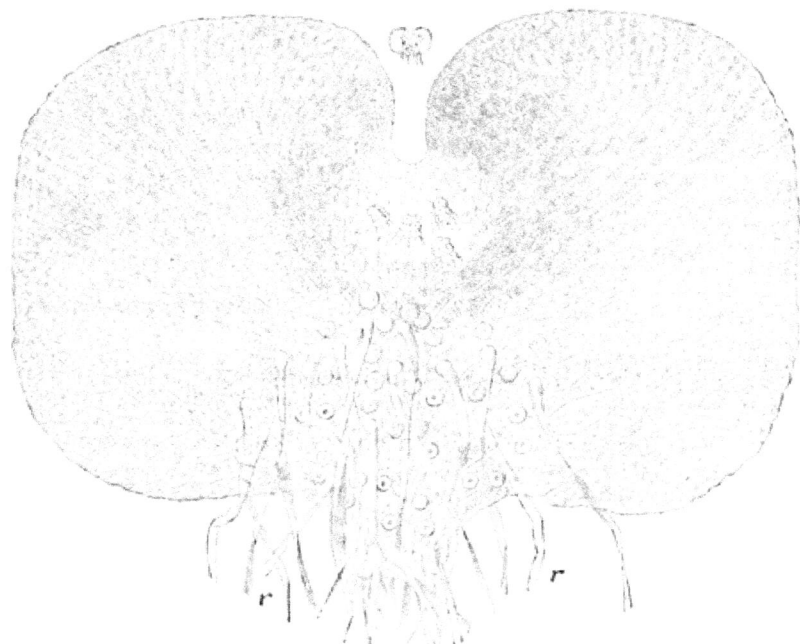

FIG. 58. (After Suminski, slightly modified.)—Adult prothallium of *Pteris serrulata* seen from below, showing the rhizoids *r* at the posterior end, the depression at the anterior end; the cushion near the latter bearing (in this case four archegonia. Amongst the rhizoids are the (spherical) antheridia. The chlorophyll-bodies only are shown in the cells of the broad plate of tissue constituting the prothallium. Just above the anterior depression is seen a prothallium of the natural size.

FIG. 59. (After Strasburger.)—Mature antheridium of *Pteris serrulata*. *p*, peripheral cells; *m*, mother-cells of the spermatozoids.

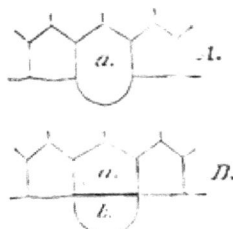

FIG. 60.—Diagram to illustrate the origin of an antheridium. *A*, very young stage; *B*, older; *a*, original epidermal cell enlarged; *b*, mother-cell of the entire antheridium.

nary cell on the lower side of the prothallium swells and forms a hemispherical or dome-shaped projection, which is soon separated by a partition from the original cell (Fig. 60). Further divisions then follow in the dome-shaped cell such that a central cell is left,

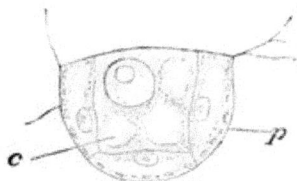

FIG. 61. (After Hofmeister.)—Later stage in the development of an antheridium of *Pteris serrulata*. *p.* peripheral cell; *c.* central cell from which the spermatozoid mother-cells arise.

surrounded by a layer of peripheral cells (Fig. 61). By repeated divisions the central cell splits up into the spermatozoid *mother-cells* (Fig. 59). Within each mother-cell the proto-plasm arranges itself in a peculiar spiral body, the *spermatozoid*, which is the *male germ-cell*.

When the mature antheridium is moistened, the peripheral cells swell and thus press out the mother-cells and spermatozoids (Fig. 62). The latter escape from the mother-cells and swim about very actively in the water. They appear as naked single cells, of a peculiar corkscrew shape, and bear upon the finer spirals numerous extremely active cilia (p. 31), by which they are driven swiftly through the water.

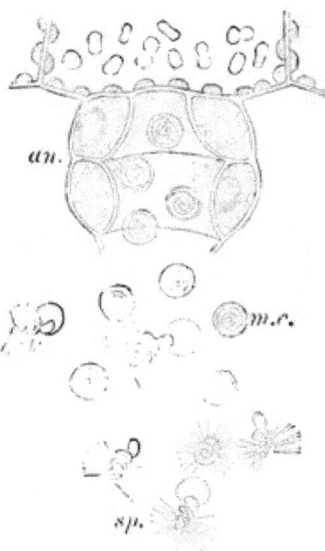

FIG. 62. (After Laerssen.)—Bursting of the antheridium and escape of the sper-matozoids. *an.* antheridium; *m.c.* sper-matozoid mother-cells; *sp.* spermato-zoids.

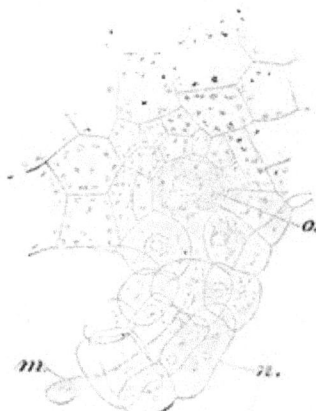

FIG. 63. (After Strasburger.)—Mature archegonium, showing the oösphere (*o.*), the neck (*n.*), and mucus (*m.*) is-suing from the mouth of the canal.

The *Archegonia*, or female organs (Figs. 58, 63), described for the first time by Suminski in 1864, likewise arise from single su-perficial cells of the prothallium. They are situated almost exclusively upon the cushion near its distal or apical extremity, and hence at the bottom of the anterior depression (Fig. 58). Since they appear later than the antheridia, they are not likely to be fertilized by spermatozoids descended from the same spore. This phenomenon of maturation of one set of sexual organs of a bisexual individual before the ripening of the other set, is a common feature among plants, and is known as *dichogamy*. There is reason to believe that important advantages are gained by thus securing cross-fertilization and preventing self-fertilization or " breeding in and in."

In the development of the *archegonium* the original cell en-larges, becomes somewhat dome-shaped, and divides by transverse partitions into three cells : a proximal, imbedded in the tissue of the prothalli-um, a middle, and a distal dome-shaped cell (Fig. 64). The fate of the proximal cell is unimportant. The

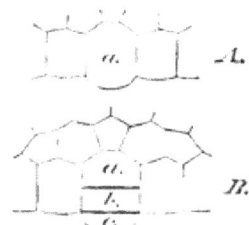

Fig. 64.—Diagram to illus-trate the origin of an arche-gonium. *A,* an early stage; *B,* a later stage; *A, a,* the original epidermal cell en-larged; *B, a,* the mother-cell; *b,* the central or canal cell; *c,* the neck-cell.

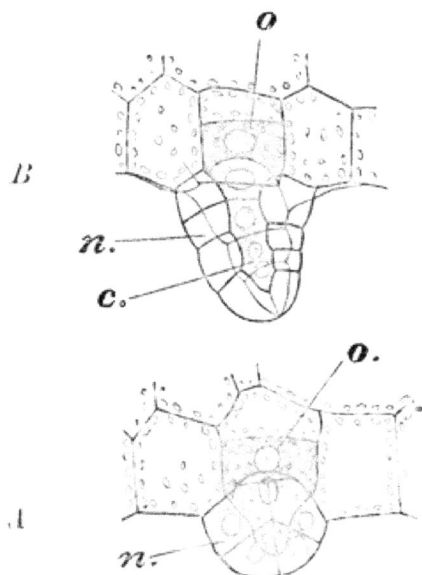

Fig. 65. (After Strasburger.)—Developing archegonia of *Pteris serrulata. A,* young stage; *B,* older; *o,* neck; *c,* canal; *o,* oösphere.

distal cell gives rise by division to a chimney-like structure, *the neck* (Figs. 63, 65), which encloses a row of cells (*canal-cells*) de-rived from the original middle cell (Figs. 63, 65). These after-wards become transformed into a mucilaginous substance filling a canal leading through the neck from the outside to the *oösphere* (Fig. 65), which also arises from the original " middle" cell at its proximal end. The oösphere is the all-important *female germ-*

7

cell to which the "neck-" and "canal-cells" are merely accessory.

Fertilization or Impregnation.

Fertilization, or the sexual act, is performed as follows : Spermatozoids in vast numbers are attracted to the mouths of the archegonia and there become entangled in the mucilage (Fig. 66). In favorable cases one or more work their way down the mucilaginous canal, and at length one penetrates and fuses with the oösphere.

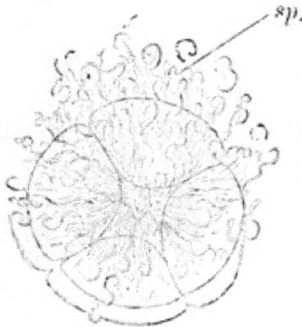

Fig. 66. (After Strasburger.)—Mouth of an archegonium of *Pteris serrulata*, crowded with spermatozoids striving to effect an entrance.

It is known that one spermatozoid is enough to fertilize the oösphere, and probably one only penetrates it; but several are often seen in the mucilaginous canal. It has recently been shown that the mucilage contains a small amount (about 0.3%) of malic acid, which probably acts both as an attraction to the spermatozoids and as a stimulus to their movements. Pfeffer has shown that capillary tubes containing a trace of a malate in solution are as attractive to the spermatozoids as the mucilage in the central canal.

The entrance of the spermatozoid into the ovum and its fusion with it mark an important epoch in the life-history of the fern. The oösphere is from this instant a new and very different thing, viz., an *embryo*, and is known as the *oöspore*. It is now the first stage of the asexual generation, though it is still maintained for some time at the expense of the sexual generation or *oöphore* (p. 89).

Growth of the Embryo.

The oöspore, or one-celled embryonic sporophore (p. 89), now rapidly becomes multicellular by dividing first into hemispheres, then into quadrants, etc. (Fig. 68; compare Fig. 21). The first plane of division is approximately a prolongation of the long axis of the archegonium (Fig. 68). The second is nearly at right angles to it, so that the quadrants may be described as anterior and posterior to the first plane. The fate of the quadrant-cells is of special importance. The lower anterior quadrant as it undergoes further division grows out into the *first root*; the upper anterior quadrant in like manner gives rise to the *rhizome* and the *first leaf.* The mass of cells derived from the two posterior quadrants remains connected with the prothallium as an organ for the absorption of nutriment from the latter, and is inappropriately called the *foot.*

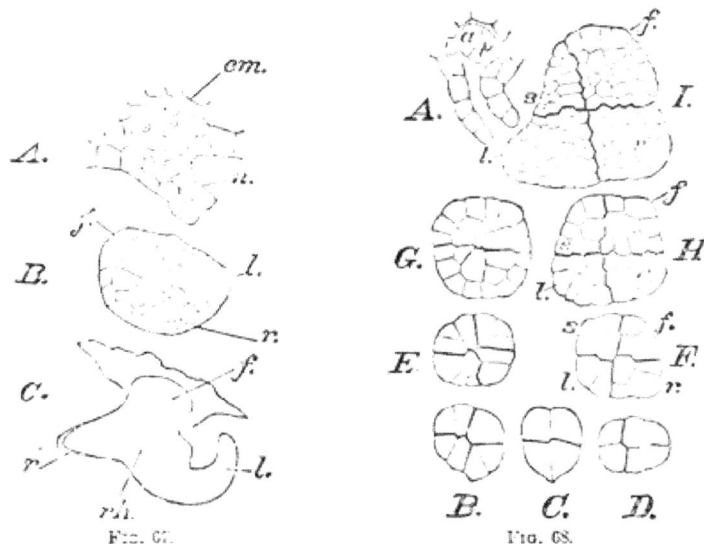

FIG. 67. (After Hofmeister.)—Development of the embryo. *A,* section showing the closed neck (*n*) and the planes of quadrant division of the oöspore or embryo (*cm*). The fore-end of the prothallium is to the right. *B* and *C,* stages of the embryo later than *A,* showing the beginnings of apical growth; *f,* foot; *l,* leaf; *r,* root; *rh,* rhizome.

FIG. 68. (From Luerssen, after Kienitz-Gerloff.)—Development of the embryo of *Pteris serrulata.* The figures are optical sections taken vertically in the antero-posterior axis of the prothallium, passing through the long axis of the neck of the archegonium; except *C* and *D,* which are taken at right angles to the others. *A, a,* and *p* are the anterior and posterior segments of the oöspore after this has divided into hemispheres. The former (*a*) forms the stem, the latter (*b*) the root. *F* shows in a late stage the division of the quadrants, *r* going to form the root, *s* the stem or rhizome, *l* the leaf, and *f* the foot; *r, l,* and *s* soon take on apical growth as indicated in *H* and *I.*

In *Pteris serrulata* the development is slightly different. The lower anterior cell becomes the first leaf; the upper anterior becomes the first portion of the rhizome, the lower posterior becomes the primary root, and the upper posterior remains as the "*foot.*"

The several parts now enter upon rapid growth accompanied by continued cell-multiplication, until a stage is reached represented in *C,* Fig. 67. A stage somewhat later than this, with its attachment to the prothallium, is shown in Fig. 69. After this the leaf grows upwards into the air, the root downwards into the earth, and the young fern begins to shift for itself. Event-

ually it reaches a condition shown in Figs. 70 and 71. The prothallium remains connected with the young fern for some time, and

may readily be found in this condition attached to flower-pots in hot-
houses, etc. But sooner or later it
falls off, and the young fern enters
upon an entirely independent existence. The appearance of the plant
and the shape of the leaf do not
always at first resemble those of the
adult fern; growth is also more
rapid at first, several leaves (7–12)
being developed successively in the
first year (p. 70).

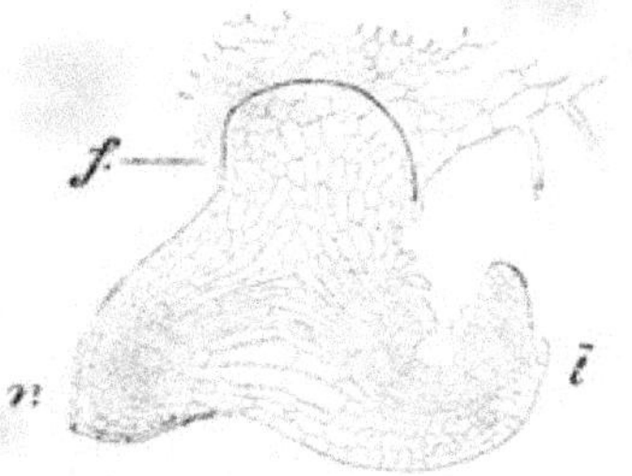

Fig. 69. (After Hofmeister.)—Young
embryo of *Pteris aquilina*, showing
its attachment to the prothallium by
the foot; *l*, leaf; *f*, foot; *r*, first root.

Differentiation of the Tissues. In the earliest stages the tissue
is nearly or quite homogeneous, i.e., meristemic. But very early
in development, as the leaf turns upwards and the root downwards, changes take place, which lead directly to a differentiation into the three great systems of tissue—epidermal, fibro-vascular, and fundamental. The epidermal and fundamental systems

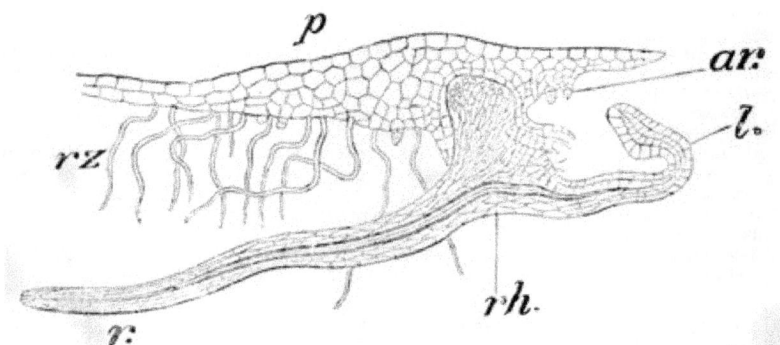

Fig. 70. (After Sachs.)—Older embryo of maiden-hair fern (*Adiantum*) attached to the
prothallium. Seen in section. *l*, leaf; *r*, first root; *rh*, beginning of the rhizome; *p*,
prothallium; *r*, rhizoids; *ar*, archegonia.

take on almost at once the peculiarities which have already been
noted in the adult, p. 75. The fibro-vascular system of tissues
is differentiated a little later. Different as the tissues of the
three systems are, it is plain from their mode of origin that all
are fundamentally of the same nature because of their descent

from the same ancestral cell; hence every cell in the plant partakes more or less completely of the nature of every other cell. The resemblances are primary and fundamental, the differences secondary and derived. And what is true of the fern in this respect is equally true of all other many-celled organisms.

Course of the Fibro-vascular Bundles.

Certain features of the disposition and course of the fibro-vascular bundles in the embryo and in the adult may conveniently be studied at this point. From the point of junction of the bundles of the first leaf and first root (Figs. 67, 68, 69) is developed one central bundle traversing the young rhizome and sending branches into the new leaves and roots until 7–9 leaves have been formed. After this time the rhizome forks, and the course of the fibro-vascular bundles in each fork is henceforwards compound. A lateral depression appears in the central bundle of each stem, rapidly increases in depth, and soon divides

FIG. 71. (After Sachs.)—Young maiden-hair fern (*Adiantum*) attached to the prothallium, *p. l*, leaf; 1, 2, the first and second roots.

the bundle into two, one upper and one lower, which are best recognized in old specimens (Fig. 35). When the forked shoots have reached a length of about three inches, these bundles send out at a small angle towards the periphery, thinner, forked branches which soon unite again to form a network near the epidermis. The uppermost of these branches, which passes in the median line above the axile bundles, is usually somewhat more fully developed, and almost as broad as the latter. This structure is generally retained in the mature rhizome (Fig. 35, *xc*). The number of peripheral bundles may be as great as twelve in the cross section. They anastomose in the vicinity of the place of insertion of each frond, and thus form a hollow, cylindrical network, having elongated meshes; but no connecting branches between them and the two axile bundles are found anywhere in the rhizome. The latter follow an entirely isolated course within the creeping stem; branches from them enter the leaves, and it is only inside the leaf-stalk that these ramifications are met by branches from the peripheral network. The bundles of the roots arise only from the peripheral bundles, but those of leaves, as already said, receive branches from both axillary and peripheral bundles. Two thick

---

* See, however, De Bary, *Comp. Anat. Phanerogams and Ferns*, p. 295. Oxford, 1884.

brown plates (*sclerotic prosenchyma*) lie between the inner and outer systems of bundles, and are only separated from one another at the sides by a narrow band of parenchyma. They are often joined on one side or even on both, in the latter case forming a tube which separates the two systems of bundles. (Hofmeister.)

Apogamy. Apospory. In rare cases, e.g., in *Pteris cretica*, the ordinary alternation of generations in the life-cycle of ferns is abbreviated by the omission of the sexual process and the immediate vegetative outgrowth of the sporophore from the prothallium (*apogamy*). In other cases there is an omission of the spore stage, and immediate vegetative development of the oöphore from the frond (*apospory*). (Cf. Farlow, *Quart. Journ. Mic. Science*, 1874; De Bary, *Botan. Zeitung*, 1878. Druery, etc.; *Journ. Royal Mic. Soc.* 1885, pp. 99 and 491.)

General Physiology.

The brake is a limited portion of matter occupying a definite position in space and time. It is bounded on all sides by material particles, some of which may be living, but most of which are lifeless. The aerial portion is immersed in and pressed upon by an invisible fluid, the atmosphere, while the underground portion is sunk in a denser medium, the earth, which likewise acts upon it. At the same time the fern reacts upon the air and the earth, maintaining during its life an equilibrium which is disturbed and finally gives way as the life of the plant draws to a close.

The Fern and its Environment. Those portions of space, earth, and air which are nearest to the brake constitute its immediate environment. But in a wider and truer sense the environment includes the whole universe outside the plant. To perceive the truth of this it is only necessary to observe how profoundly and directly the plant is affected by rays of light which travel to it from the sun over a distance of many millions of miles, or how extremely sensitive it is to the alternations of day and night or of summer and winter. The plant is fitted to make certain exchanges with its environment, drawing from it certain forms of matter and energy, and returning to it matter and energy in other forms. Its whole life is an unconscious struggle to wrest from the environment the means of subsistence; death and decay mark its final and unconditional surrender.

In its relations to the environment the fern illustrates an important and fundamental biological law, viz., that *the living organism must be adapted to its environment*, or, in other words, that a certain *harmony* between organism and environment is essential to the continuance of life, and any influence which tends to disturb or destroy this harmony tends to disturb or destroy life. Thus the aerial part of *Pteris* must be fitted to make exchanges with, and maintain its life in, the atmosphere, while the underground part must be similarly "adapted" to the soil in which it lives and with which it has to deal.

At the same time it appears that the general environment of the plant considered as a whole falls naturally into secondary or minor environments, to which special portions of the plant must be adapted. Hence arises the differentiation of parts fitted to cope with the various minor environments. This necessity extends to every organ and tissue of the plant, each of which has its peculiar function corresponding to its special environment.

The aerial part of *Pteris* displays admirable adaptation to, or harmony with, the environment in its stalk, which rises to a point of vantage for procuring air and light, and in its broadly spreading top, which is covered by a skin, tough and impervious to prevent undue evaporation and consequent desiccation, yet translucent, to allow the sun's rays to reach the starch-making tissue within. The rhizome also, with its pointed terminal buds, its elongated roots, armed with boring tips, and its thick, fleshy parenchyma for the storage of food, is well adapted to, and in harmony with, its own special environment. In order to realize this, we have only to imagine the fern to be inverted, the aerial portion being planted in the earth, and the underground portion lifted into the air and exposed to the winds and sunshine. Under these circumstances the want of adaptation of the parts to their respective environments would speedily become apparent. Yet different as these parts now are, they have lately sprung from the same cell. More recently they were barely distinguishable in a mass of tissue, part of which turned upwards into the air, while another part turned downwards into the earth. But as development went on, the aerial and underground parts were progressively differentiated, thus becoming more and more perfectly adapted to the peculiar conditions by which each is surrounded.

Thus it appears that the harmony between every part of the organism and its environment is brought about by a *gradual process* in the history of each individual plant. And there is every reason to believe that this is typical of the history of each *species* of plant or animal in the past. We know that environments change, and that to a certain extent organisms change correspondingly, provided the change of environment be not too sudden or extreme. The environment of *Pteris* changes periodically with the regular alternation of summer and winter, and the plant also undergoes a corresponding periodic change of structure in order to maintain its adaptation to the environment. During the summer the aerial part is fully developed, and, as a result of its activity, starch is accumulated in the rhizome. At the approach of winter the aerial part dies, and the plant is reduced to the underground part safely buried in the soil. During the winter and spring the starch is gradually consumed, and the aerial part is put forth again as the aerial environment becomes once more favorable to it. The organism, therefore, possesses a certain *plasticity* which enables it to adapt itself to gradually changing conditions of the environment.

A little consideration will show that every function or action of living things may be regarded as contributing to the same great end: viz., harmony with the environment; and from this point of view life itself has been defined as "*the continuous adjustment of internal relations to external relations.*" *

Now there is good reason to believe that as environments have gradually undergone extensive changes in the past, organisms have gradually undergone corresponding changes of structure. Those which have become so modified as to be most perfectly adapted to the changed environment, have tended to survive and leave similarly adapted descendants. Those which have been less perfectly adapted have tended to die out through lack of fitness for the environment; and by this process—called by Spencer the "Survival of the Fittest" and by Darwin "Natural Selection"—the remarkable adaptations everywhere met with have been gradually worked out.

Nutrition. The fern does work. In pushing its stem through the soil, in lifting its leaves into the air, in moving food-matters from point to point, in building new tissue, in the process of re-

---

* Spencer, *Principles of Biology*, Vol. I., p. 80. N. Y., Appleton, 1884.

production, and in all other forms of vital action, the plant uses force and loses energy. In doing any kind of work, portions of the organism are consumed and cast out as burnt or waste materials which mostly take the form of carbon dioxide and water; and these phenomena go on to a certain extent whether the plant feeds or not. But it is evident that these losses of energy and of matter must ultimately be made good by fresh supplies entering the plant, and the income must equal the outgo if the fern is merely to hold its own, and must exceed it if the fern is to grow.

Thus it comes about that there is a more or less steady flow of matter and of energy through the living organism, which is itself a centre of activity, like a whirlpool (p. 2). The chemical phenomena accompanying the flow of matter and energy through the organism are those of *nutrition* in the widest sense. This term is more often restricted especially to the phenomena accompanying the income, while those pertaining to the outgo are regarded as belonging to *excretion*. The intermediate processes directly connected with the life of protoplasm are put together under the head of *metabolism*, though this word is sometimes used in a broader sense to designate collectively all of the chemical phenomena of protoplasm.

The Income. The income of the fern is of two kinds, viz., matter and energy. Matter enters the plant in the liquid or gaseous form by *diffusion*, both from the soil through the roots (liquids), and from the atmosphere through the leaves (gases). Energy may enter the plant to a small extent as the potential energy of food, but comes in principally as the kinetic energy of sunlight absorbed in the leaves. The table on p. 106 shows the precise nature and the more important sources of the income.

Of these substances the solids (salts, etc.) must be dissolved in water before they can be taken in. Water and dissolved salts continually pass by diffusion from the soil into the roots, where together they constitute the sap. The sap travels throughout the whole plant, a main though not the only cause of movement being the constant *transpiration* (evaporation) of watery vapor from the leaves, especially through the stomata. The gaseous matters (carbon dioxide, oxygen, nitrogen?) enter the plant mainly by diffusion from the atmosphere, are dissolved by the sap in the leaves, and thus may pass to every portion of the plant.

Pteris owes its power of absorbing the energy of sunlight to

INCOME OF *PTERIS.*

MATTER.	WHENCE DERIVED.
Carbon.	Mainly from the atmosphere as carbon dioxide (CO_2).
Hydrogen.	Mainly from the soil as water (H_2O).
Oxygen.	Mainly from the soil as water (H_2O) and from the air as free oxygen.
Nitrogen.	Mainly from the soil * as nitrates or ammonium compounds.
Sulphur.	Mainly from the soil as sulphates.
Other elements.	Mainly from the soil as various salts.
ENERGY.	
Kinetic.	Mainly from the sunlight through the leaves.
Potential.	Perhaps to a limited extent in food materials *via* the roots.

the *chlorophyll-bodies;* for plants which, like fungi, etc., are devoid of chlorophyll are unable thus to acquire energy. Entering the chlorophyll-bodies, the kinetic energy of sunlight is applied to the decomposition of carbon dioxide (CO_2) and water (H_2O). After passing through manifold but imperfectly known processes, the elements of these substances finally reappear as starch ($C_6H_{10}O_5$) in the form of granules imbedded in the chlorophyll-bodies, and free oxygen (O), most of which is returned to the atmosphere. Thus the leaf of *Pteris* in the light is continually absorbing carbon dioxide and giving forth free oxygen.

Carbon dioxide and water contain no potential energy, since the affinities of their constituent elements are completely satisfied (cf. p. 33, Chap. III.). Starch ($C_6H_{10}O_5$), however, contains potential energy, since the molecule is relatively unstable, i.e., capable of decomposition into simpler, stabler molecules in which stronger affinities are satisfied. And this is due to the fact that in the manufacture of starch in the chlorophyll-bodies the kinetic energy of sunlight was expended in lifting the atoms into

* It is generally stated that plants are unable to make use of free atmospheric nitrogen, but recent investigations have thrown doubts upon this view. (Cf. Atwater, *American Chemical Journal*, Vol. VI. [1885] p. 365.)

a position of vantage, thus endowing them with energy of position. In this way some of the radiant and kinetic energy of the sun comes to be *stored up* as potential energy in the starch. In short, *Pteris*, like all green plants, is able by co-operation with sunlight to use simple compounds, poor in energy or devoid of it, and out of them to manufacture *food*, i.e., complex compounds rich in potential energy. We shall see hereafter that this power is possessed by green plants alone; all other organisms being dependent for energy upon the potential energy of ready-made food. This must in the first instance be provided for them by green plants; and hence without chlorophyll-bearing plants animals (and colorless plants as well) could not long exist.

The plant absorbs also a small amount of kinetic energy, independently of the sunlight, in the form of heat; this, however, is probably not a source of vital energy, but only contributes to the maintenance of the bodily temperature (cf. p. 40).

Metabolism. Starch, as has just been seen, is first formed in the chlorophyll-bodies. But the formation of starch, all-important as it is, is after all only the manufacture of food as a preliminary to the real processes of nutrition. These processes must take place everywhere in ordinary protoplasm; for it is here that oxidations occur and the need for a renewal of matter and energy consequently arises (cf. pp. 33 and 34). Sooner or later the starch grains are changed into a kind of sugar (*glucose*, $C_6H_{12}O_6$), which, unlike starch, dissolves in the sap, and may thus be easily transported to all parts of the plant. Wherever there is need for new protoplasm, whether to repair previous waste or to supply materials for growth, the elements of the starch (or glucose) are in some unknown way combined with nitrogen and sulphur (probably also with salts, water, etc.), to form proteid matter. The particles of this newly formed compound are incorporated into the protoplasm (by "intus-susception," p. 5) and, in some way at present shrouded in mystery, are endowed with the properties of life. We do not know how long they may remain in the living state, but sooner or later they are oxidized, and as a result of the oxidation, that energy is set free which enables the fern to do work and prolong its existence. The oxidated products are afterwards eliminated from the cells.

If a larger quantity of starch is formed in the chlorophyll-bodies than is immediately needed by the protoplasm for pur-

poses of repair or growth, it may be re-converted into starch
after journeying as glucose through the plant, and be laid down
as "reserve starch" in the parenchyma of the rhizome, or elsewhere. When this reserve supply is finally needed at any point
in the plant, it is again changed to glucose and transported
thither. It is probable that new leaves and new tissues generally, are always formed in part from this reserve starch, and
not solely from newly-formed starch.

The Outgo. The outgo, like the income, is of two kinds, matter and energy, but it cannot be so readily tabulated.

The plant suffers annually a great loss both of matter and of
potential energy in the production of spores and in the autumnal
dying-down of the leaves. But matter also leaves the plant daily
as carbon dioxide (in small quantities), water, and oxygen, both by
diffusion through the epidermis and by transpiration through the
stomata. Strictly speaking, the term outgo should be restricted
to the output of matter which has at some time actually formed
a part of the living protoplasm, and this does not apply to the
oxygen, which is simply given off in the manufacture of starch,
or to the bulk of the water of transpiration which passes straight
through the plant without undergoing any chemical change. Energy likewise leaves the plant continuously both as *heat* and in
the *doing of mechanical work*, both of which are involved in
every vital act.

Respiration. It has been remarked that in the light (i.e., when
manufacturing starch) *Pteris* takes in carbon dioxide and gives off
free oxygen. But if the plant be deprived of light, as at night, the
reverse is true, and the plant takes in a small amount of oxygen
and gives off a corresponding amount of carbon dioxide.
This latter process is the true *breathing* or *respiration* of the
plant, and it must not be confounded with the taking in of
carbon dioxide and giving off of oxygen. Respiration goes on in
the light also, probably with greater energy than in darkness, but
it is then partially obscured by the other and more conspicuous
process.

We have seen that energy is set free in living matter by a
decomposition of its own substance which is really a process of
oxidation or combustion, in which free oxygen plays an important part (p. 34, Chap. III.); hence the absorption of free oxygen in respiration. Among the products of the combustion,

water and carbon dioxide are the most important; and this is the origin of the carbon dioxide given off. It will appear hereafter that precisely the same action takes place in the respiration of animals, and that all living things breathe or respire in essentially the same way.

It was for a long time believed that a leading difference between plants and animals lay in the fact that the former give off oxygen and absorb carbon dioxide, while the latter give off carbon dioxide and absorb oxygen. But it is now known that both give off carbon dioxide and require oxygen, and that only the chlorophyll-bearing parts of green plants are endowed with the special function of decomposing carbon dioxide and water, and manufacturing starch—as a result of which they do (but in the light only) give off oxygen.

PTERIS AQUILINA.

(Balance-Sheet of Nutrition.)

INCOME.	OUTGO.
Matter.	*Matter.*
Food,	Carbon dioxide,
Inorganic salts,	Water,
Carbon dioxide,	Excreted substances,
Water,	Reproductive germs,
Free oxygen.	Leaves, etc.,
	Free oxygen—from decomposition of carbon dioxide in light.
Energy.	*Energy.*
Sunlight absorbed by chlorophyll,	Work performed,
Heat,	Heat,
(Potential energy in foods?)	Potential energy in cast-off matters, reproductive germs, etc.

Balance in favor of *Pteris:*

Matter.
  Tissues, protoplasm, starch, cellulose, chlorophyll, etc.
Energy.
  Potential energy in organic matters.

Physiology of the Tissue-Systems.

The *epidermal* tissues serve as the sole medium of exchange between the inner parts of the plant and the environment; they are protective, and in certain regions are useful for support. The function of reproduction also falls upon these tissues, as is shown by the development of the sporangia, antheridia, and archegonia.

The *fibro-vascular* tissues serve in part as a supporting skeleton, for which functions their richness in prosenchyma and their firm continuity admirably adapt them. An equally important function, however, is their *conductivity*, since they serve for the transportation (especially) of the water for evaporation by the leaf (*transpiration*), and for the carriage (through the sieve-tubes) of the undissolved and indiffusible proteids. The *funda-mental tissues* are devoted either to sharing the special duties of the other systems, as in the case of the sclerotic parenchyma abutting upon the epidermal tissue in the rhizome (p. 77), and the sclerotic prosenchyma which appears to behave like the fibro-vascular tissues; or to nutritive and metabolic functions, as in the mesophyll (p. 86) and the parenchyma of the rhizome.

Reproduction. It is not known whether the brake ever dies of old age. Barring accidents, growth at the apical buds seems to be unlimited, keeping pace with death of the hinder parts of the rhizome (p. 69). But whether the individual dies or not, ample provision against the death of the race is made in the function of reproduction. Although reproduction is useless to the individual and even entails upon it serious annual losses of matter and energy, yet to this function every part of the plant directly or indirectly contributes. The reproductive germs are carefully prepared; are provided with a stock of food sufficient for the earliest stages of development; and are endowed with the peculiar powers and limitations of *Pteris aquilina*, which influence their life-history at every step and are by them transmitted in turn to their descendants. In short, reproduction is the supreme function of the plant. If we may paraphrase the words of Michael Foster, the oösphere is the goal of individual existence, and life is a cycle beginning with the oösphere and continually coming round to it again.

PRACTICAL STUDY.

A. **Introductory.**

1. If possible, visit some place where *Pteris aquilina* grows abundantly. Study it *in situ*, observing closely the nature of its immediate environment, whether wet or dry, shady or sunny, etc., and the nature of the soil. Dig up an entire plant, following the ramifications of the rhizome, and observing its depth in the soil, connections with the leaves, etc.

2. Observe in a *whole plant* the form and size; aerial and underground parts; axis and appendages; rhizome and branches;

roots; apical buds; decaying ends; mature and embryonic leaves; the stumps of older leaves, etc.

3. Examine the *leaves* and make out all their different parts in small and large specimens (p. 68). Observe in fertile leaflets the reflexed edges (*indusia*) and the sporangia.

B. Anatomy of the Rhizome and the Stipe.

1. Cut a piece of the mature rhizome squarely and smoothly across, and in an end view observe with the naked eye or a hand-lens (Figs. 33, 35)—

 a. The general outline.
 b. Lateral ridges.
 c. Epidermis.
 d. Fibro-vascular system.
 e. Fundamental system, including all that remains, and
   showing,
    α. Parenchyma.
    β. Sclerotic parenchyma or sub-epidermis.
    γ. Sclerotic prosenchyma.
    δ. The insertion of a root (occasionally).

2. Make a similar cut ¼ inch from the end through an apical bud, compare the end view of this with what you have seen in the mature rhizome, and make sketches.

3. Examine a thick longitudinal section through an apical bud. Afterwards soak the section for a few minutes in a solution of aniline chloride acidified with hydrochloric acid. All of the tissues except the fundamental parenchyma are colored yellow. Sketch.

4. Dissect with a knife a piece of a mature rhizome softened by soaking for some hours in water, or by heating for a few minutes in Schulze's macerating fluid (see Appendix), and follow out the various tissues.

5. Make a similar examination of the *stipe*.

C. Histology of the Rhizome.

1. Using a sharp razor or a microtome, make an entire transverse section of a small mature rhizome. If an alcoholic specimen is used, it should be previously soaked for some minutes in

water, and kept wet during the cutting. Mount the section in
water, and, using a low power, observe the general arrangement
of the three great TISSUE-SYSTEMS (p. 72), and compare Figs. 33,
35-41.

2. In a thin transverse section study with a higher power the
component tissues and cells of each system, as follows :—

    a. EPIDERMAL SYSTEM.

    b. FUNDAMENTAL SYSTEM.
        α. Fundamental parenchyma.
        β. Sclerotic parenchyma.
        γ. Sclerotic prosenchyma.

    c. FIBRO-VASCULAR SYSTEM.
        α. Bundle-sheath.
        β. Phloëm-sheath.
        γ. Bast-fibres or phloëm-prosenchyma.
        δ. Sieve-tubes.
        ε. Phloëm-parenchyma.
        ζ. Tracheids.
        η. Vessels.
        θ. Wood-parenchyma.

3. Repeat 1 and 2, using instead of a mature part of the
rhizome a section taken ¼ inch from the tip of an apical bud.
Compare these sections with the former and report your results.

4. With a knife-blade or a "lifter" transfer a thin section to
a solution of iodine in a watch-glass. Turn the section once or
twice to insure its thorough staining ; transfer after a few minutes
to a larger dish of water to remove the excess of iodine : mount
in water, or dilute glycerine (glycerine 1 part, water 2 parts), and
study carefully with your highest power the distribution of
starch and protoplasm in the several systems, tissues, and cells.

Prepare tables showing (*a*) the contents of the tissues and (*b*)
their relative vitality or activity as indicated by their contents
and the nature of their walls.

5. Stain other sections in acidulated solution of aniline chloride, mount as before, and study. (Cf. B. 3.)

6. Make a transverse section of a *root*, and compare it with
that of the rhizome.

7. Prepare thin longitudinal sections of the mature rhizome, imbedding in celloidin or between pieces of pith, according to instructions by the demonstrator (see Appendix). Stain some of them in iodine, and others in aniline chloride as before. Mount a number of sections in dilute glycerine, and study carefully, comparing with the cross-sections. Make out the three systems, and the constituent tissues and cells.

8. Repeat the foregoing observations with thin, longitudinal sections passing through an apical bud.

D. Meristem. Apical Cell. Differentiation.

    1. In good sections prepared as in C. 8, observe—

        a. The general shape of the bud, the funnel-shaped pit at the extremity, and the hairs which cover it.

        b. Search for the apical cell, comparing Figs. 42–44.

        c. Notice the undifferentiated condition of the tissue (*meristem*) at the extremity, its richness in protoplasm, and the progressive differentiation behind the apical cell.

        d. Repeat these studies upon transverse section, sketching especially the developing cells of sclerotic prosenchyma and the tracheids.

        e. Stain a number of sections with borax carmine, iodine solution, aniline chloride, eosine and methyl green, and compare carefully

E. Histology of the Aerial Parts.

    1. The Stipe.

Study transverse and longitudinal sections of the stipe. Sketch.

    2. The Lamina.

        a. Moisten a leaflet with water, scrape lightly the upper surface, and mount in water the transparent colorless bits of *epidermis* thus peeled off. Sketch.

        b. Stain with iodine, and report your results.

        c. Scrape and examine the lower surface as in (*a*). Be careful to get only colorless bits. Study, then stain with iodine and sketch. In the stomata notice especially the guard-cells, their contents, and the chink (*stoma*).

8

d. Scrape as before, this time examining the green cells (*mesophyll*) from the upper and the lower parts. Compare them and explain their differences.

e. Study a bit of epidermis taken from the under side where it covers a veinlet. Stain with iodine and aniline chloride, and see if the cells agree chemically with the ordinary epidermal cells from the top or the bottom of the leaflet.

f. Imbed a bit of a leaflet in celloidin or in pith and make cross-sections. (It is difficult to get thin sections of *Pteris*, but with many hot-house ferns it is easier.) Mount in glycerine and observe—

  α. Epidermis.

  β. Mesophyll.

  γ. Stomata.

  δ. Fibro-vascular bundles in the midrib and veinlets.

  ε. Prosenchyma, above and below the fibro-vascular bundle of the midrib which it helps to form.

  ζ. Indusium and *reproductive organs* when present (Fig. 52).

  η. Hairs, when present.

g. Study the venation of a leaflet (Fig. 50) and compare it if possible with that of *Osmunda regalis.*

F. Reproduction.

  1. SPORANGIA.

  a. In fertile leaflets dissect off the indusium and study the *sporangia,* the receptacle, etc., with a hand-lens or the naked eye.

  b. Mount sporangia of different ages in water or dilute glycerine, and study them first with a low and then with a high power. Observe the *capsule, pedicel, annulus, spores* (or *archesporium*), etc., and make sketches.

  2. SPORES.

  Study spores of different ages and observe their contents. Crush a mass of them, stain with iodine, and report your results.

  3. GERMINATION.

  Fill a porous vessel (a flower-pot will answer) with clean fine sand. Set the vessel in a shallow dish of water, and when

the surface of the sand is moist sow upon it fern-spores provided
by the demonstrator. Cover the whole apparatus with a bell-
glass, and after three or four days remove some of the surface
layer of sand to a slide, add a drop of water, and examine with a
low power. If none of the spores have burst, repeat the operation every week till some are found in germination. Follow this
process, making sketches of the different stages.

4. PROTHALLIUM.

Prothallia are common in most hot-houses, especially in the
fern-house. They grow on the sides and the soil of flower-pots,
on the earth under the benches, etc.

> *a.* Examine a prothallium with the naked eye or a hand-
> lens, then mount it in water and examine with your
> lowest power. Sketch.
>
> *b.* Turn it over and sketch again.
>
> *c.* With your highest objective, study some of the mar-
> ginal cells, observing the protoplasm, chlorophyll-
> bodies, etc.

5. THE SEXUAL ORGANS.

Find the cushion and search for the sexual organs. Observe
also the rhizoids and their superficial (epidermal) origin.

> *a. Antheridia.* Observe the size, shape, and contents of
> antheridia of different ages, crush them, and observe
> more carefully the mother-cells. If possible, find
> moving spermatozoids.
>
> *b. Archegonia.* On the same or another prothallium
> study the archegonia. Then cut sections of the
> prothallium between pieces of moist cork or pith,
> and make out the details of their structure. Stras-
> burger recommends the piling of several prothallia
> (carefully freed from sand) one upon another before
> cutting, thus permitting a number of sections to be
> cut at once.

6. FERTILIZATION.

This is not easy to observe, but the attempt may be made by
examining successively a number of very fresh prothallia in different stages. They must be mounted carefully (not flooded
with water), and are generally easier to observe after the specimen
has been mounted a little while.

7. Embryology

Except in its general features, this is too difficult for the beginner. He may, however, observe the later stages by studying old prothallia with the young fern just appearing, and young ferns with the prothallia still adherent.

G. Chlorophyll and Starch.

Vigorous prothallia afford excellent examples of cells bearing chlorophyll-bodies in which starch is easily detected.

1. Study some of the marginal cells with your highest power, paying special attention to the chlorophyll-bodies and their arrangement.

2. In favorable cases observe the opaque rod-like or oval starch grains inside them.

3. Prove by reagents that these are starch grains.

CHAPTER VII.

THE BIOLOGY OF AN ANIMAL

THE COMMON EARTHWORM.

(*Lumbricus agricola*, Hoffmeister; *L. terrestris*, Linnæus.)

WE have now to study the biology of some animal which will introduce us to those structures and functions which are characteristic of animals generally, and will afford a basis for comparison with the plant. Any higher animal would serve this purpose, but the common earthworm is a peculiarly favorable object of study, because of the simplicity of its structure, the clearness of its relation to other animals standing above and below it in the scale of organization, and the ease with which it may be procured and dissected. Earthworms, of which there are many species, are found in all parts of the world, extending even to isolated oceanic islands. In the United States there are several species, of which *Lumbricus agricola* is much the most common. This is the familiar earthworm, which has long been known but was first scientifically described under the name *Lumbricus terrestris* by Linnæus more than a hundred years ago. It was afterwards found necessary to change the name to *L. agricola*.

Mode of Life, etc. Earthworms live in the earth, burrowing through the soil at a depth varying from a few inches to several feet. Here they pass the day-time, crawling out at night or after a shower. The burrows proceed at first straight downwards, and then wind about irregularly, sometimes reaching a depth of six or eight feet. The earthworm is a nocturnal animal, and during the day lies quiet in its burrow near the surface, extended at full length, head uppermost. At night it becomes very active, and, thrusting the fore end of the body far out, explores the vicinity in all directions, still clinging fast, however, to the mouth of the burrow by the hinder end. In this way the worm is able to forage, seizing leaves, pebbles, and other small objects, and dragging them into the burrow. Some of these are devoured; the remainder (including the pebbles, etc.) are used to line the upper part of the burrow, and to plug up its opening when the worm retires for the day. Besides bits of leaves and animal matter, earthworms swallow large quantities of earth, which is passed slowly through the alimentary canal, so that any nutritious substances contained in it may be digested and absorbed. This earth is generally swallowed at a considerable distance below the surface of the ground, and is finally voided at the surface near the opening of the burrow. In this way arise the small piles of earth ("*castings*" or *faces*) which every one has seen, especially in the morning, wherever earthworms abound. Very large quantities of earth are thus brought to the surface by earthworms —in some cases, according to Darwin's estimates, more than eighteen tons per acre in a single year. In fact, most soils are continually being worked over by worms; and Darwin has shown that these humble creatures, in the course of centuries, have helped to bury huge rocks and the ruins of ancient buildings.*

The earthworm has no ears, eyes, or any other organs of special sense. Nevertheless—and this is a point of great physiological interest—the fore end of the body is sensitive to light; for if a strong light be suddenly flashed upon this part of the worm as it lies stretched forth, it will often "dash like a rabbit into its burrow." Of touch the animal has a keen sense, as may be proved by tickling it; and its sense of taste must be well developed, since the worm is somewhat fastidious in its choice of food. Earthworms appear to be quite deaf, but possess a distinct, though feeble, sense of smell.

General Morphology.

Attention will first be directed to certain features of the body seemingly of little importance, but really full of meaning when compared with like features in other animals higher or lower in the scale of organization.

Antero-posterior Differentiation. The body (Fig. 72) has an elongated cylindrical form, tapering to a blunt point at

* Darwin, *Vegetable Mould and Earthworms.* Appleton, N. Y., 1882.

one end, obtusely rounded and flattened at
the other. As a rule, the pointed end
moves forwards in locomotion, and the
mouth opens near it. For these and other
reasons the pointed end might be called the
head-end, and the other the tail-end. But
the worm has really neither head nor tail,
and hence the two ends may better be distinguished as the *fore-end* and the *hinder-
end*, or still better as *anterior* and *posterior*.
And in scientific language the fact that the
worm has anterior and posterior ends which
differ from each other is stated by saying
that it shows *antero-posterior differentiation*. This simple fact acquires great importance in the light of comparative zoölogy; for it may be shown that the antero-
posterior differentiation of the earthworm,
insignificant as it seems, is only the beginning of a series of important modifications
extending upwards through more and more
complex stages to culminate in man himself.

Dorso-ventral Differentiation. In living
or well-preserved specimens, the body is not
perfectly cylindrical, but is somewhat flattened, particularly near the posterior end,
and has a slightly prismatic four-sided form.
One of the flattened sides, slightly darker
in color than the other, is habitually turned
upwards, and is therefore called the back,
the opposite or lower side, commonly turned
downwards, being the belly. For the sake
of accuracy, however, biologists are wont to
speak of the *dorsal aspect* (back) and *ventral
aspect* (belly) of the body; and the fact that

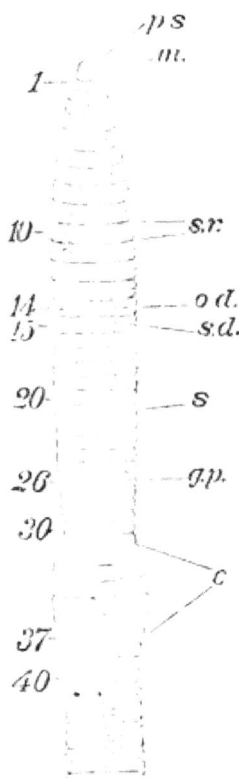

FIG. 72.—Enlarged view of the anterior and posterior parts
of the body of an earthworm as seen from the ventral aspect. *oa*, anus; *c*, clitellum; *g.p.*, glandular prominences on
the 26th somite; *m*, mouth; *o.d.*, external openings of the
oviducts; *p.s.*, prostomium; *s*, setæ; *s.r.*, openings of the
seminal receptacles; *s.d.*, external openings of the sperm-
ducts. (Slightly enlarged.)

an animal has a back and belly differing from each other in structure or function, or both, as in the earthworm, is expressed by saying that the body exhibits *dorso-ventral differentiation*. This, like antero-posterior differentiation, is very feebly expressed in the external features, though clearly marked in the arrangement of the internal parts of the earthworm. In higher animals it becomes one of the most conspicuous features of the body.

Bilateral Symmetry. When the body is placed in the natural position, with the ventral aspect downwards, a vertical plane passing longitudinally through the middle will divide it into exactly similar right and left halves. This similarity is called two-sidedness or *bilateral symmetry.* Though not very obvious externally, this symmetry characterizes the arrangement of all the internal parts; and it may be gradually traced upwards in higher animals, until it becomes as striking and perfect as in the human body.

Thus a very superficial examination reveals in the earthworm two fundamental laws of organization, viz., *differentiation* or the law of difference, and *symmetry* or the law of likeness. And these laws are of interest for the reason among many others that earthworms, like other organisms, have had a history as a race, have *come to be* by a gradual process (cf. p. 104). And biology must answer the questions *how* and *why* certain parts have become symmetrical and others differentiated. Without entering into a full discussion of the question at this point, it may be said that the main cause of symmetry or differentiation has probably been likeness or unlikeness of function or of relation to the environment. Earthworms show antero-posterior and dorso-ventral differentiation, because the anterior and posterior extremities, or the dorsal and ventral aspects, have been differently used and exposed to different conditions of environment. And on the other hand, the organism is bilaterally symmetrical, because the two sides have been similarly used and have been exposed to like conditions of environment.

Metamerism. Another general feature of the earthworm is of great importance in other animals, including the higher forms. The body is marked off by transverse grooves into a series of similar parts like the joints of a bamboo fishing-rod, or like the joints of fingers (Fig. 72). These parts are called *metameres*, or more often *somites*, and the body is consequently said to have a *metameric* structure, or to exhibit *metamerism*. From the outside, the somites appear to be produced simply by regular folds in the skin, like the wrinkles between the joints of our fingers. But as the wrinkles of the fingers are only the external expression of a more fundamental jointed structure within, so the external folds separating the somites represent an internal division into successive parts, which affects all the organs of the body, and is a result of some of the most important phenomena of development.

The explanation of metamerism, or "*serial symmetry*," is not so easy as that of bilateral symmetry; it is in fact one of the most difficult problems of morphology. But it will be seen farther on that metamerism, so clearly and simply expressed in the earthworm, can be traced upward in ever-increasing complexity to the highest forms of life, and suggests some of the most interesting and fundamental problems with which biology—and especially morphology—has to deal. Indeed, the comparative study of the anatomy of most higher animals consists very largely in tracing out the manifold transformations of their complicated somites, which under many disguises can be recognized as fundamentally like the simpler somites of the earthworm.

Modifications of the Somites. The somites differ considerably in different parts of the body. The extreme anterior end is formed by a smoothly rounded knob called the *prostomium*, which is shown by its mode of development not to be a true somite. It forms a kind of overhanging upper lip to the *mouth*, which lies just behind it on the ventral aspect. Behind the mouth is the first somite, in the form of a ring,* interrupted above by a backward prolongation of the prostomium.

The somites from the 1st to the 27th are rather broad, and gradually increase in size. Between the 7th and 19th somites are a variable number, which are often swollen on the ventral side, forming there *glandular prominences* by which the worms adhere to each other during copulation. Between the 25th and 35th (the number and position varying slightly in different specimens) the somites are swollen above and on the sides, and the folds between them are scarcely defined except on the ventral aspect. Taken together, they form a broad, conspicuous girdle called the *clitellum* (Fig. 72, *c*), whose function is to secrete the

---

* In numbering the somites the prostomium must never be reckoned, the first somite being *behind the mouth*.

capsule in which the eggs are laid, and also a nutritive milk-like fluid for the use of the developing embryos. Behind the clitellum the somites are less broad, somewhat four-sided in cross-section, and flattened from above downwards. This flattening sometimes becomes very conspicuous towards the posterior end. Towards the very last they decrease in size rather abruptly, and end in the *anal* somite, which is perforated by a vertical slit, the *anus* (Fig. 72. *an*). All the somites are perforated by small openings leading into the interior of the body, and forming the outlet for numerous organs: their position will be described in treating of these organs (see also the practical study, p. 168). Each somite, excepting the anterior two or three and the last, gives insertion to four groups of short and minute bristles or *setæ*, which are arranged in four longitudinal rows along the body. Two of these rows run along the ventral aspect, two are near the sides. The setæ extend outwards from the interior of the body, where

Fig. 73.—Diagram to illustrate the action of the setæ. The dotted outline represents the position of the seta and its muscles when bent in the opposite direction. *m*, muscles; *s*, seta; *w*, body-wall.

they are supplied with small muscles by which they can be turned somewhat either forwards or backwards, and can also be protruded or withdrawn (Fig. 73). The setæ are of great use in locomotion. When pointed backwards they hold the worm fast as it crawls forwards: when turned forwards the worm creeps backwards. They are of interest therefore as representing an extremely simple and primitive form of limb.

Plan of the Body. The body of the earthworm (Fig. 74), like

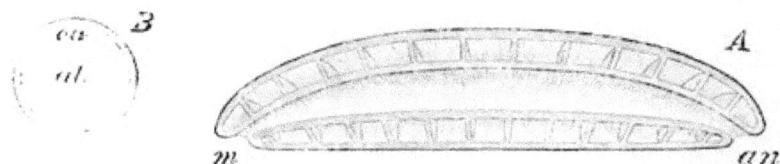

Fig. 74.—*A*, diagram of the earthworm as seen in a longitudinal section of the body, showing the two tubes, the cœlom, and the dissepiments. *B*, diagram of cross-section. *al*, alimentary tube; *an*, anus; *co*, cœlom; *m*, mouth.

that of all higher animals, consists of two tubes, one (*al*) within the other and separated from it by a considerable space or cavity (*ca*). The inner tube is the alimentary canal, opening in front by the *mouth* and behind by the *anus;* the outer tube is the body-wall, and its cavity is the *body-cavity* or *cœlom.* The cœlom is not, however, a free continuous space extending from end to end, but is divided transversely by a series of thin muscular partitions, the *dissepiments*, into a series of closed chambers traversed by the alimentary canal. Each compartment corresponds to one somite, the dissepiments being opposite the external furrows mentioned on p. 120. All the organs of the body are originally developed from the walls of these chambers, and some of them (e.g., the organs of excretion) project into the cavities of the chambers, that is, into the cœlom.

Organs and Systems. The apparatus or organs by which the worm lives and makes exchanges with its environment are widely different from those of the fern; and since they are more complex, and from a mechanical point of view more perfect, the animal is said to be *more highly organized.* In the fern certain tissues were found to be constantly associated in definite groups or *systems* (fundamental, epidermal, and fibro-vascular systems, p. 72). In *Lumbricus* like associations or systems exist, each consisting of a number of different *organs* which co-operate to perform some important function, such as alimentation, circulation, reproduction, etc.

Such groups of co-operative organs are known as *organic systems*, and they take their special names from the leading functions—alimentation, respiration, excretion, etc.—which they respectively perform. They are shown in the following table :—

ANIMAL FUNCTIONS AND SYSTEMS OF ORGANS.

a. Alimentary.		
b. Circulatory.		
c. Metabolic.	**Nutrition.**	
d. Excretory.		
e. Respiratory.		For the Individual.
f. Motor.		
g. Co-ordinative.		
h. Sensitive.	**Relation.**	
i. Supporting, Connecting, Protecting, etc.		
j. Reproductive.		For the Race.

Since living matter undergoes incessant waste of material by oxidation (p. 4), and is continually expending energy (p. 35), the earthworm must be replenished with matter and energy in the form of ordinary food (*Alimentation*) and oxygen (*Respiration* in part). The new matter must be distributed to all parts of the organism (*Circulation*), and after its incorporation into the living substance and ultimate decomposition for the liberation of energy (*Metabolism*) the results of decomposition must be eliminated from the body (*Excretion; Respiration* in part). Again, every animal possesses some power of *Motion* either as a whole or in some of its parts. The movements as well as all the other vital actions must be guided and controlled (*Co-ordination*). The organism must be capable of receiving impressions from the external world, and of perceiving changes in or near itself (*Sensation* in the broadest sense). And besides the above-mentioned requirements there are a number of subsidiary needs, such as the support and protection of the soft and delicate structures of the body; means of offence, defence, etc. (*Support, Connection, Protection*, etc.). Lastly, since individual existence is limited, provision must be made for the perpetuation of the race (*Reproduction*).

These ten primary functions fall at once naturally into two groups, since the first nine relate to the existence of each individual, while the tenth concerns the individual only as a member of the race of earthworms. The nine individual needs may in turn be arranged in two groups, viz., **Nutrition** and **Relation**. The former (Alimentation, Circulation, Metabolism, Respiration, and Excretion) are concerned especially with the income, transformation, and outgo of matter and energy in the organism. The latter (Motion, Co-ordination, Sensation, Support, etc.) have to do especially with the mechanical relations of the various parts of the organism to one another and to the environment.

The machinery by whose action the primary functions are performed varies widely in different animals even for accomplishing the same function. For instance, the single function of Respiration is performed by four quite different mechanical arrangements in earthworms, beetles, fishes, and men, respectively. Locomotion (a special case of motion) is effected by different mechanisms, according as it occurs in earthworms, jelly-fishes, butterflies, or horses.

Sometimes the machinery assumes the form of simple cells, as in the corpuscles of the blood; sometimes the form of simple tissues, as in the epidermis, which protects the free surfaces of the body. Very often it appears as distinct organs, like the heart or the lungs. But as a rule the mechanism consists of a number of different tissues or organs (or both), which work together for a common end, thus constituting a system.

SPECIAL MORPHOLOGY.

Alimentary System. This consists of the inner body-tube or *alimentary canal*, which is divisible into several differently constructed portions playing different parts in the process of alimentation. Going backwards from the mouth these are as follows:—

1. The *pharynx* (Fig. 75, *ph*), an elongated barrel-shaped pouch extending to about the 6th somite. Its walls are thick and muscular, and from their cœlomic surface numerous small muscles radiate on every side to the body-wall. When these muscles contract, the cavity of the pharynx is expanded; and if the mouth has been previously applied to any solid object, such as a leaf or pebble, the pharynx acts upon it like a suction-pump. In this way the animal lays hold of the various objects, nutritious and otherwise, which it devours or draws into its burrow.

2. The *œsophagus* (*œ*), a slender, thin-walled tube extending from the 6th to the 15th somite. Through this the food is swallowed, being driven slowly along by wave-like (*peristaltic*) contractions (p. 129). In the region of the 11th and 12th somites are three pairs of small pouches opening at the sides of the œsophagus. These are the *calciferous glands* (*c.gl*). They contain solid masses of calcium carbonate, and Darwin conjectures that their use is partly to aid digestion by neutralizing the acids generated during the digestion of leaves, and perhaps partly to serve as an outlet for the excess of lime in the body, especially when worms live in calcareous soil.

3. The *crop* (*c*), about the 16th somite; a thin-walled, sac-like dilatation of the alimentary canal, which serves as a reservoir to receive the swallowed food.

4. The *gizzard* (*g*), about the 17th somite; a cylindrical, firm and muscular portion, lined by a horny membrane. In this the food is rolled about, squeezed and ground to prepare it for digestion.

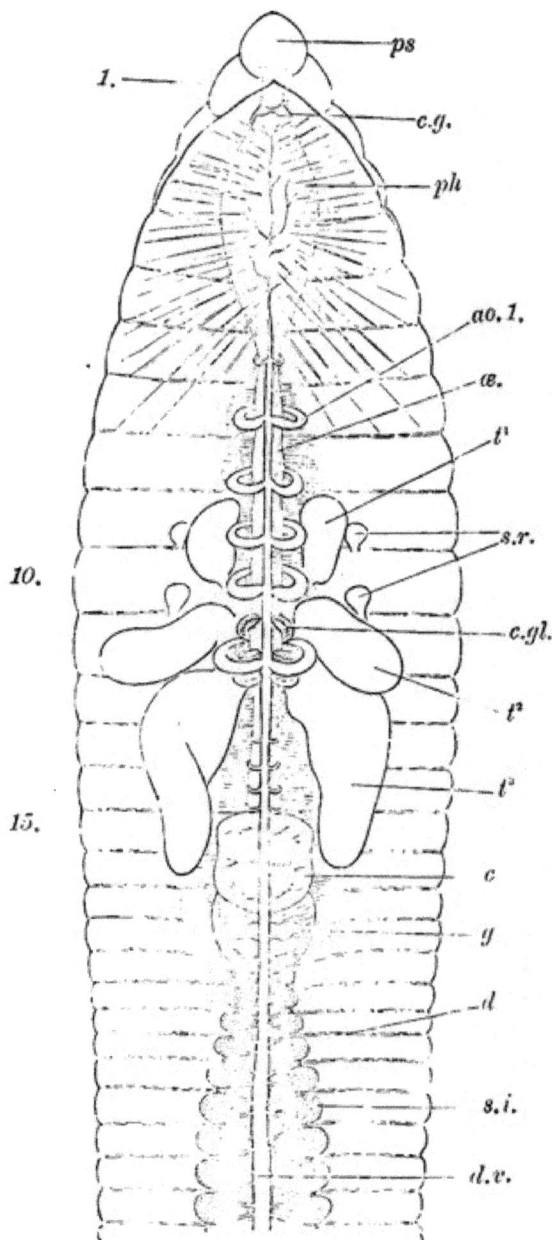

FIG. 75.—Dorsal view of the anterior part of the body of *Lumbricus*, as it appears when laid open along the dorsal aspect. *ao*, aortic arch; *c*, crop; *c.g.* cerebral ganglia; *c.gl*, calciferous glands; *d*, dissepiment; *d.v*, dorsal vessel; *g*, gizzard; *œ*, œsophagus; *ph*, pharynx; *ps*, prostomium; *s.i*, stomach-intestine, showing the lateral pouches ; *s.r*, seminal receptacles; *t¹*, *t²*, *t³*, testes.

5. The *stomach-intestine* (*s.i*), corresponding to both the stomach and intestine of higher animals. It is a straight, thin-walled tube extending from the gizzard to the anus. As food passes along this tube it is first digested in it, then absorbed from it; and finally the residue of refuse matters is voided at the anus. We cannot distinguish any dividing-line, however, between the region where the food is principally digested (*stomach*) and the part where it is principally absorbed (*intestine*). Probably both processes go on along its entire length, but digestion, as might be expected, appears to be carried on especially in the anterior part.

The surfaces of the stomach-intestine are largely increased by the presence of lateral pouches or diverticula, one on either side in each somite. In front these are large and conspicuous, but behind they gradually diminish in size until scarcely perceptible.

The inner surface of the stomach-intestine is further increased by a deep inward fold, called the *typhlosole*, running longitudinally along the dorsal median line. The typhlosole is not visible on the exterior, but is seen by opening the stomach-intestine from the side or below, or upon making a cross-section. The outer surface of the stomach-intestine is covered above, at the sides, and within the typhlosole, by yellowish-brown cells, which are most abundant in the anterior portion. These cells are believed to be of a glandular nature, their office being to separate from the blood certain substances which are of use in digestion. They have been compared to the liver of certain animals, but probably the comparison is not very accurate. Hence the non-committal term "*chloragogue cells*" has been proposed and is used for them.

Circulatory System. The distribution of absorbed food-matters is effected in the earthworm by two different devices.

1. *Cœlomic.* The cavity of the cœlom is filled with a colorless fluid which must be regarded as a kind of blood, and the dissepiments are perforated by irregular openings, and the cœlomic fluid is probably driven back and forth through all parts of the cœlom. As the digested food is absorbed from the stomach-intestine a considerable part of it is believed to pass into the cœlomic fluid, and is thus conveyed directly to the organs which this fluid bathes. The cœlomic fluid is composed of two constituents, viz., a colorless fluid called the *plasma*, and colorless isolated cells or *corpuscles* which float in the plasma, and are remarkable for the fact that they undergo constant, though slow,

changes of form. In fact they closely resemble certain kinds of *Amœba*, and we should certainly consider them to be such if we found them occurring free in stagnant water. We know, however, that they live only in the *plasma*, and for this and other reasons we must regard them not as individual animals, but as constituent cells of the earthworm. The cœlomic fluid is in fact a kind of *tissue* consisting of isolated colorless cells floating in a fluid intercellular substance.

2. *Pseudhæmal.* Besides the cœlomic circulation there is another and more complicated circulatory apparatus consisting of tubes branching throughout the body. Through these tubes is driven a red fluid formerly supposed to correspond to the red blood of higher animals. It is now known, however, that this fluid is only in part comparable to true blood ; hence it is called *pseudhæmal* fluid (i.e., *false blood*), and the system of tubes is called the *pseudhæmal* system. Like the cœlomic fluid, the pseudhæmal fluid contains corpuscles ; but these are extremely difficult to observe, since their index of refraction is very nearly the same as that of the fluid.[*]

The pseudhæmal tubes (which for the sake of brevity may be called *blood-vessels*) ramify throughout the whole body, permeating nearly all the tissues and organs, and forming an exceedingly complicated system. Amidst all this complication we can, however, readily distinguish certain general features which are of the greatest importance when compared with the blood-vessels of higher animals. These are as follows :—

　　a. The *dorsal vessel* (Fig. 75, *d.v*), a long muscular tube lying along the upper side of the alimentary canal. In the living worm it may be distinctly seen through the semi-transparent skin as a dark-red band, which is tolerably straight when the worm is extended, but is made zigzag during contraction. If it be closely observed, a sort of wave-like contraction is often seen running from behind forwards. This may be very clearly observed in a worm stupefied by chloroform, especially if it has been laid open along the dorsal side. The dorsal vessel then appears as a

---

[*] We shall see hereafter that the red blood of higher animals corresponds to the *cœlomic* and *pseudhæmal* fluids taken together.

deep-red, somewhat twisted tube running along the
upper side of the alimentary canal. Wave-like contractions continually start from its hinder end and
run rapidly forwards, one after another, to the anterior end, where the dorsal vessel finally breaks up
on the pharynx into a large number of branches
(Fig. 75).

The result of these orderly progressive contractions is that
the fluid within the tube is driven forwards—very much as the
fluid in a rubber tube is forced along when the tube is stripped
through the fingers. This action of the vessels is known as *peristaltic contraction.*

The peristaltic contraction of the pseudhaemal vessel, unlike that of the
rubber tube, is intrinsic and automatic (p. 32), being caused by the contraction of its own muscular walls. The physiological importance of peristaltic contractions is very great. For in animals generally the blood is
propelled in this way; the contents of the alimentary canal are carried on-
wards and finally driven out; urine is ejected, and the products of reproduction (eggs, embryos, etc.), as well as the male elements, are similarly
transported or expelled.

b. Sub-intestinal vessel. This is a straight vessel which
runs along the middle line on the *lower* side of the
alimentary canal, parallel to the one just described.
It returns to the hinder part of the body the fluid
which has been carried forwards by the dorsal vessel.
On the pharynx it breaks up into many branches,
which receive the fluid from corresponding branches
of the dorsal vessel.

3. Circular vessels, which run from the dorsal vessel downwards around the alimentary canal to join the sub-intestinal vessel. Through these vessels the fluid flows down to join the back-
ward-flowing current in the sub-intestinal vessel, thus completing
a circuit. Of these circular vessels there are usually two pairs
(sometimes three or four) in each somite, partly imbedded in the
wall of the alimentary canal and covered with "chloragogue
cells" (p. 127), so that their red color is not apparent. In front
of the 13th somite, however, there is only one pair in each
somite. The five pairs lying between the 7th and the 11th
somites (inclusive) are greatly enlarged, and quite free from the

9

wall of the œsophagus (Fig. 75, *ao*). These vessels pulsate regularly like the dorsal vessel, powerful peristaltic contractions continually passing along them from above downwards. Owing to their great size and powerful pulsations they must be a very efficient means of maintaining the flow of the pseudhæmal fluid; in fact they are often called the "*hearts*" of the earthworm. It is misleading, however, to call them hearts, because they do not correspond exactly to the hearts of higher animals. The earthworm has no true heart, but only a system of pulsating bloodvessels. Instead of "hearts," these five pairs of circular vessels should be called the *aortic arches*.

The pseudhæmal fluid makes in its flow a circuit of the whole body, passing forwards in the dorsal vessel, downwards in the anterior circular vessels (especially in the aortic arches), backwards in the sub-intestinal vessel, and then up into the dorsal again by vessels in the hinder part of the body. Inasmuch as the flow makes this complete *circuit*, it is spoken of as a *circulation* in the strict sense of the word. From the main vessels numerous branches are given off which ramify through every part—on the walls of the body and alimentary canal, and over and through all of the organs contained in the cœlom.

In the foregoing account only the more obvious features of the pseudhæmal vessels have been mentioned, and many important details have been passed over. The circular vessels of the stomach-intestine can be followed for only a short distance out from the dorsal vessel, where they seem to break up into a large number of small parallel vessels lying close together and running around to the lower side. Here they do not directly join the sub-intestinal, but empty into a sinus or vessel which runs parallel to the latter, closely imbedded in the wall of the stomach-intestine. The sub-intestinal vessel proper is quite separate from the stomach-intestine, and communicates by short branches (usually two in each somite) with the vessel lying above it. This may be clearly seen in the region of the gizzard. On this there is a variable number of small lateral vessels, which break up partly into a branching network, and are partly resolved into extremely fine parallel vessels surrounding the organ. On the crop are three or four pairs of lateral branches from the dorsal vessel which branch out into a fine network, but do not break up into parallel vessels as on the gizzard. In the two somites (13th and 14th) in front of the crop there are usually two pairs of vessels running around the œsophagus. In the 11th and 12th somites a small branch is given off to each calciferous gland. The most anterior pair of circular vessels are in the 6th somite, and are very small. In front of this the dorsal vessel breaks up into the pharyngeal network. In front of the 11th somite there are three sub-intestinal vessels, as shown in Fig. 76. The two additional vessels lie on either side of the primary one and break up into branches at the sides of the pharynx. The aortic arches empty into the middle vessel, and at the point of junction there is a communication with the lateral vessel of the corresponding side.

Besides the dorsal and sub-intestinal vessels there are three other minor longitudinal trunks (Fig. 77). Two of these are very small, and lie on either side above the nerve-cord (p. 136), sending fine branches out from

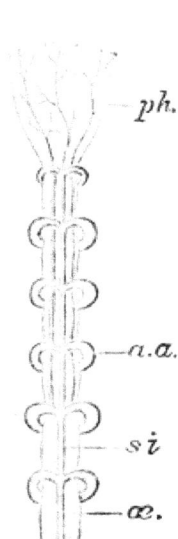

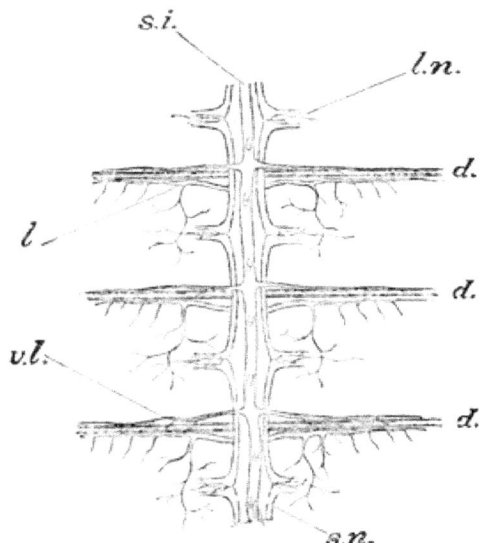

FIG. 76. FIG. 77.

FIG. 76.—Ventral view of part of the pharynx and œsophagus, showing the three sub-intestinal vessels in this region, the first pair of circular vessels, and the aortic arches. *a.a*, aortic arches; *œ*, œsophagus; *ph*, pharynx; *s.i*, sub-intestinal vessel.

FIG. 77.—A portion of the ventral nerve-chain seen from above, and showing some of the blood-vessels; *d.d*, dissepiments; *l*, lateral vessel from the sub-neural vessel; *l.n*, lateral nerves proceeding from a ganglion; *s.i*, sub-intestinal vessel, showing along its upper side the cut ends of short vessels which connect it with the sinus running along the ventral side of the alimentary canal; *s.n*, supra-neural vessel; *v.l*, ventro-lateral vessel.

each ganglion along the lateral nerves. These are the *supra-neural* trunks (*s.n*).

The third longitudinal vessel (*sub-neural* trunk) lies below the nerve-cord. From it a pair of lateral vessels are given off in each somite just behind the dissepiment (*l*, Fig. 77). A large branch from this vessel (often more than one) runs backward to ramify on the body-wall. The main trunk runs outward in the dissepiment and gives off numerous branches to the adjoining nephridium (p. 132).

Besides the lateral vessels from the sub-neural and supra-neural there are two other sets. These are (1) the *dorso-lateral* vessels, given off from the dorsal vessel on each side in each somite on the anterior face of the dissepiment, and running downwards and outwards; and (2) the *ventro-lateral* vessels (*v.l.* Fig. 77), which occupy a corresponding position on the lower side, arising from the sub-intestinal vessel. Thus there are four sets of *lateral* vessels arising from the *dorsal*, *sub-intestinal*, *supra-neural*, and *sub-neural* trunks, respectively. All run outwards, either in the dissepiments or on the inner face of the body-wall, and break up into a most complicated capillary network, which covers the entire inner face of the body-wall, ramifying through its substance and through all the organs of the body. The exact course of the blood in this network has not been determined.

Metabolic System. See p. 164.

Excretory System. It is the office of the excretory system to remove from the organism the waste matters resulting from the breaking down of living tissue. This does not mean the passing

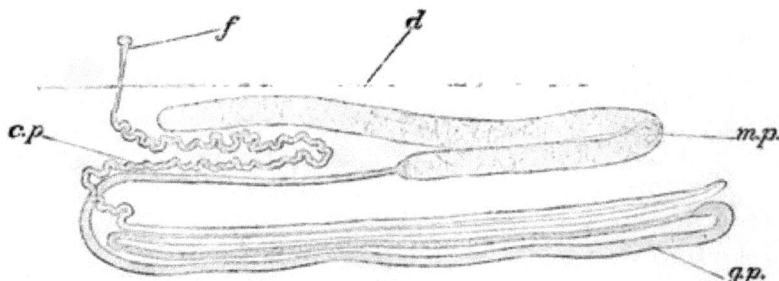

Fig. 78.—An entire nephridium with its folds somewhat separated for the sake of clearness. *c.p*, ciliated part; *d*, dissepiment; *f*, funnel at the inner opening; *g.p*, glandular part; *m.p*, muscular part.

away of indigestible materials through the anus, for such matters have never been absorbed and built up into tissue. Excretion means the removal from the body of matter which has really formed a part of its substance, but has been used up and is no longer alive. In our own bodies this function is performed chiefly by the kidneys and the skin, the waste matters passing off in urine and sweat. In the earthworm it is principally performed by small organs called *nephridia*, of which there are two in each somite, excepting the first three or four (Fig. 80).

Each nephridium (Fig. 78) consists of a long convoluted tube, attached to the hinder face of a dissepiment, and lying in the cœlom at the side of the alimentary canal. At one end the tube passes through the body-wall and opens to the exterior by a minute pore situated between the outer and inner rows of setæ (p. 122.) The other end of the tube passes through the dissepi-ment very near to the point where this is penetrated by the nerve-cord (p. 136), and opens by a broad, funnel-like expansion into the cavity of the next somite in front (*f*, Fig. 78). The margins of the funnel and the inner surface of the upper part of the tube are densely covered with pow-erful cilia (Fig. 79), whose action tends to produce a current setting from the cœlom into the funnel and through the nephridium to the exterior.

The nephridium is covered with a net-work of blood-vessels, and its walls con-tain glandular cells which have the power of removing from the blood a substance (*urea*) which is the most important con-

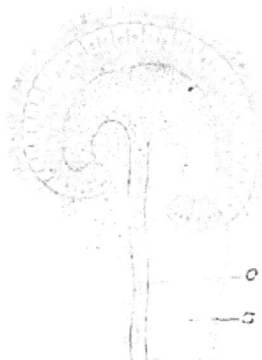

FIG. 79.—A nephridial funnel much enlarged, showing the cilia, the beginning of the cil-iated canal (*c*), and the outer sheath (*s*).

stituent of the urine of higher animals. This substance is passed into the cavity of the nephridium, and so out to the ex-terior.

Several distinct parts may be distinguished in the nephridium, as follows :—

1 The *funnel;* this is flattened with an oblique opening, and is com-posed of beautiful ciliated columnar cells set like a palisade around its edge. It leads into—

2. The *ciliated (uriniferous) tubule (c.p.* Fig. 78), which is a very long, thin-walled tube bending back and forth and lined by long active cilia. This suddenly enlarges to join—

3. The *glandular* part (*g.p*): a stouter thick-walled tube, whose walls contain numerous granular gland-cells in which the urea is probably secreted. This enlarges in turn to join—

4. The *muscular* part (*m.p*); much larger and shorter than the pre-ceding, its walls containing numerous muscular fibres crossing one another in all directions. This part forms a kind of sac or reservoir like a bladder, in which the excreted matter may accumulate and be passed out to the exterior.

The various parts of the nephridium are held together by connective tissue (p. 155), and are covered with a rich network of blood-vessels.

Respiratory System. Respiration or breathing is a twofold operation, consisting of the taking in of free oxygen and the giving off of carbon dioxide, by gaseous diffusion through the surface of the body. Strictly speaking, this free oxygen must be regarded as food, while carbon dioxide is to be regarded as one of the excretions. Hence respiration is tributary both to alimentation and excretion; but since many animals possess special mechanisms to carry on respiration, it is convenient and customary to treat of it as a distinct process effected by a distinct respiratory system.

Respiration in this sense is essentially an exchange of gases between the blood and the air, carried on through a delicate membrane lying between them. The earthworm represents the simplest conditions possible, since the exchange takes place all over the body, precisely as in a plant. Its moist and delicate walls are everywhere traversed by a fine network of blood-vessels lying just beneath the surface. The oxygen of the air, either in the atmosphere or dissolved in water, readily diffuses into the blood at all points, and carbon dioxide makes its exit in the reverse direction. Freed of carbon dioxide and enriched with oxygen, the blood is then carried away by the circulation to the inner parts, where it gives up its oxygen to the tissues and becomes once more laden with carbon dioxide.

It is interesting to study the various devices by which this function is performed in different animals. In the earthworm the whole outer surface is respiratory, and no *special* respiratory organs exist. In other animals these arise simply by the differentiation of certain regions of the general surface, which then carry on the gaseous exchange for the whole organism. In many aquatic animals such regions bear filaments or flat plates or feathery processes known as *gills* or *branchia*, which are bathed by the water containing dissolved air, though in many such animals respiration takes place to some extent over the general surface as well. In insects the respiratory surface is confined to narrow tubes (*tracheæ*) which grow into the body from the surface and branch through every part, but must nevertheless be regarded as an infolded part of the outer surface. In man and other air-breathing vertebrates the respiratory surface is mainly confined to the lungs, which are simply localized infoldings of the outer surface specially adapted to effect a rapid exchange of gases between the blood and the air.

Moreover, it is easy to see why special regions of the outer surface have in higher animals been set aside for respiration.

It is essential to rapid diffusion that the respiratory surface should be covered with a thin, moist membrane, and it is no less essential that many animals should be provided with a firm outer covering as a protection against mechanical injury or desiccation. Hence the outer surface becomes more or less distinctly differentiated into two parts, viz., a protecting part, the general integument; and a respiratory part, which is usually preserved from injury by being folded into the interior as in the case of lungs or tracheae, or by being covered with folds of skin as in the gills of fishes, lobsters, etc.

Motor System. The movements of the earthworm as a whole (locomotion) are produced by the action of muscles forming part of the wall of the body. But besides these muscles there are many others, of which the most important are those which move the setae (p. 122), the muscles of the dissepiments (p. 159) and of the pharynx (p. 125), the muscles of the nerve-cord and those by which the various peristaltic contractions are performed, the most noteworthy being those of the alimentary canal, of the blood-vessels, and of the nephridia. All of these muscles belong to the motor system, which, therefore, does not form an isolated system, but is intimately connected with many others.

The muscles of the body-wall are arranged in two concentric layers below the skin. In the outer layer the muscles run around the body, and are therefore called *circular* muscles. Those of the inner layers have a *longitudinal* course,—i.e., parallel with the long axis of the body,—and are arranged in a number of different bands. The most important of these are:

1. The *dorsal band* (see Fig. 87), covering the entire back, and extending as far down on either side as the outer row of setae.

2. The *ventral bands*, covering the ventral side between the two lower (i.e., inner) rows of setae. These vary greatly in different parts of the body, and in some parts there are additional narrow bands.

The setae, which must be reckoned as part of the motor system, are produced by glandular cells covering their inner ends, and they grow constantly from this point, somewhat as hairs grow from the root. After being fully formed, and after a certain amount of use, the setae are cast off and replaced by new ones which have meanwhile been forming. In each group we find therefore setae of different sizes. At their inner ends they are covered by a common investment of glandular cells which appears as a slight rounded prominence when viewed from within. These prominences are called the *setigerous glands*. When a worm is laid open from above the glands are seen in four parallel rows, two of which lie on either side of the nerve-cord (see Fig. 80).

Nervous System. Co-ordinating Mechanism. By means of the nervous system, the various parts of the body are co-ordinated or made to work in harmony. Through its agency also the worm receives impressions from the environment in the form of sensations (sight, smell, touch, etc.) which enable it to act according to the requirements of its daily life, or, in other words, to maintain harmony with the environment.

These functions are performed by slender cords of protoplasm, the *nerves*, which receive stimuli and conduct impulses in an orderly manner from point to point through the whole organism; and by *nerve-cells* (centres), which receive, co-ordinate, and send forth the impulses carried by the nerves. The nerve-cells are arranged in small masses or *ganglia*, which, taken together, constitute the *central nervous system;* the nerves collectively constitute the *peripheral nervous system.*

1. *Central Nervous System.* This consists in the earthworm of two parallel cords of nervous tissue running along the middle line of the body on the ventral side, just inside the body-wall and beneath the sub-intestinal blood-vessel (Fig. 80). Throughout the greater part of the body the two cords are so closely united as to appear like one, the double character only appearing in cross-section (p. 160). At about the 4th somite, however, the two cords separate completely, and embrace the front part of the pharynx. Each cord in this region runs obliquely forwards and upwards at the side of the pharynx and joins its fellow on the upper side, forming a complete ring around the pharynx, the *pharyngeal collar* (*p.c.* Fig. 80).

Behind this the double cord is usually called the *ventral nerve-chain.* About the middle of each somite it is somewhat enlarged to form a distinct *ganglion* (*e.g.* Fig. 80). The ventral cord is, in fact, a chain of such ganglia connected by double cords called *commissures.*

Besides the ventral ganglia, there are two very distinct ganglia lying side by side on the upper surface of the pharynx where the two halves of the pharyngeal collar join (Fig. 80, *c.g*). These are the *cerebral ganglia,* so called because they are supposed to correspond, in a general way, with the brain (*cerebrum*) of higher animals.

Besides these main parts of the central system, there are many smaller

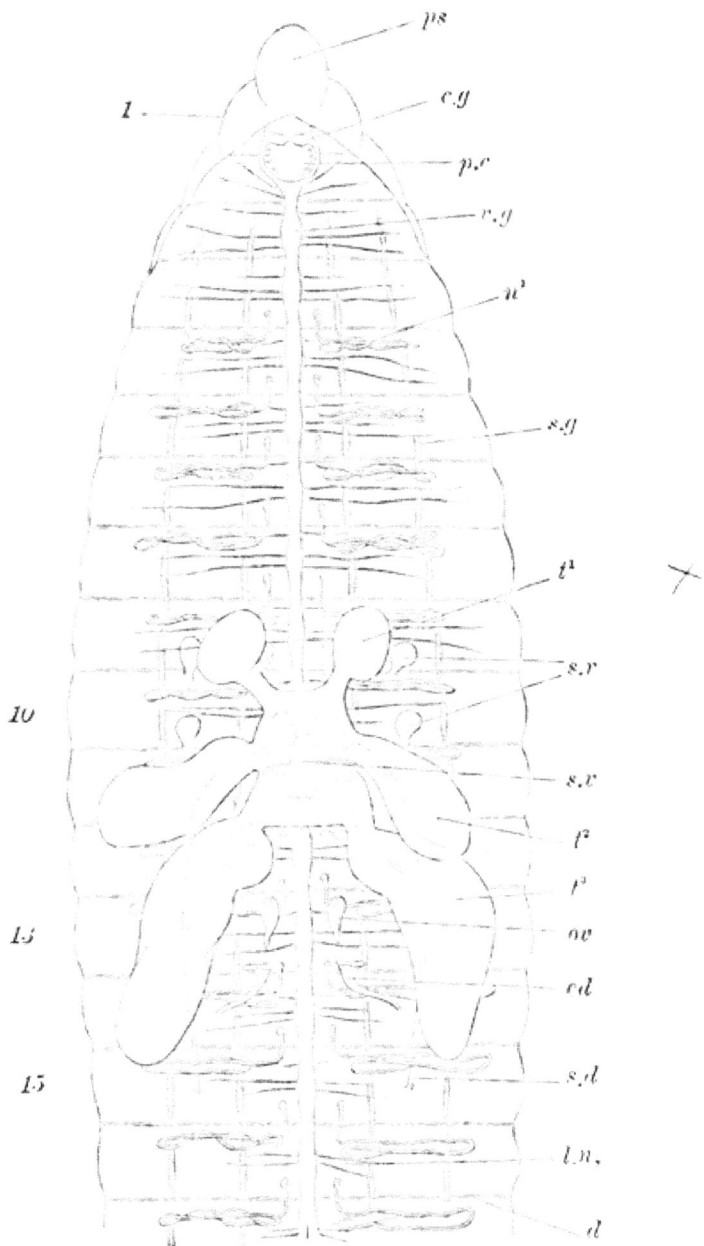

FIG. 80.—Anterior portion of an earthworm laid open from above, with the alimentary and circulatory systems dissected away. *c.g*, cerebral ganglia; *d*, dissepiment; *l.n*, lateral nerves; *n¹*, first nephridium; *od*, oviduct; *ov*, ovary; *p.c*, pharyngeal collar, inside which are the pharyngeal ganglia; *ps*, prostomium; *s.d*, sperm duct; *s.g*, setigerous glands; *s.r*, seminal receptacles; *s.v*, seminal vesicle; *t¹, t², t³*, testes; *v.g*, ventral ganglion.

ganglia in various parts of the body. Of these the most important are the *pharyngeal ganglia.*—3 to 5 in number—which lie on the wall of the pharynx on each side just within the pharyngeal collar. They are connected with the latter by fine branches, and send minute nerves out upon the walls of the pharynx. This series of ganglia is often inappropriately called the *sympathetic* system.

Peripheral Nervous System. To and from the central system just described run the nerves which constitute the peripheral system. These are as follows :—

1. *A pair of nerves* running out on either side of each ventral ganglion and lost to view among the muscles of the body-wall (*l.n.* Fig. 80).

2. *A single nerve* proceeding from the ventral commissures on each side immediately behind the dissepiment to which it is mainly distributed.

3. *A pair of nerves* (Fig. 80), in front of the foremost pair of ventral ganglia.

4. *A pair of nerves* from the point where the two ventral cords diverge in front of the next dissepiment to form the pharyngeal collar.

5. *A nerve* from each half of the collar just beyond its divergence from its fellow. (Origin incorrectly shown.)

6. *Two large nerves* (*cerebral*), which run forwards from the cerebral ganglia, break up into many branches, and are distributed to the anterior part of the body.

The branches of the various nerves ramify through all parts of the body. Many of them—notably those of the cerebral ganglia—end in the skin, and conduct impulses inwards to the nerve-centres (*afferent* nerves). Some end in the muscles, and conduct impulses outwards to them from nerve centres (*efferent* or *motor* nerves). Others terminate in the walls of the stomach, nephridia, blood-vessels, etc., and among these are probably both afferent and efferent nerves. It is highly probable, from analogy with higher animals, that the afferent and efferent nerve-fibres are always distinct—i.e., that the inward and outward impulses never travel through the same nerve-fibre.

Nervous Action. Nerves are normally thrown into action by certain *changes* or *stimuli* affecting their extremities. Thus a touch or pressure upon the skin acts as a stimulus to the nerves ending near the point touched. The nerves may be stimulated by a great variety of agents:—by mechanical disturbance, as in the case just cited, by heat, electricity, chemical action, and in special cases by waves of light or of sound. They may also be stimulated by vital changes taking place in the nerve-cells or other structures in which they terminate. Whatever be the nature of the stimulus, the effect is always the same, viz., the origination of a *nervous impulse* which travels rapidly through the nerve to its opposite end, and there either throws into action the organ connected with it, or modifies actions already occurring. The nature of the nervous impulse is very imperfectly understood, but it seems to be a progressive molecular movement, which may be roughly compared to an electric current passing through a wire.

By means of this delicate mechanism every part of the body is placed in material connection with, and is able to influence the activity of, many other parts. And thus, like the soldiers of an army, all the organs are enabled to act together as a unit, performing different functions by a physiological division of labor, yet co-operating for the common welfare.

The actions of the different organs are for the most part co-ordinated by a process known as *reflex action*, which plays a highly important part in the lives of all higher animals. It is illustrated by the diagram, Fig. 81.

A and *B* are two organs (or other structures) whose action is to be co-ordinated by means of the nervous system. They are not directly connected by nerves, but both are connected with a nerve-centre, *C*, which is a nerve-cell or group of nerve-cells situated in one of the ganglia. If *A* be thrown into action, an *afferent impulse* travels to *C*, excites the nerve-centre, and causes an *efferent* impulse to travel out to *B*, which is thereby thrown into action also, or is modified in respect to actions already going on. Thus the actions of *A* and *B* are co-

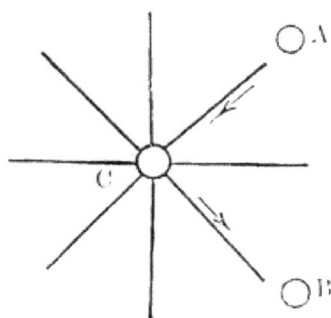

Fig. 81.—Diagram to illustrate reflex action; *A* and *B*, two organs; *C*, a nerve-centre; the paths of the impulses are indicated by the arrows.

ordinated through the agency of *C*; the whole chain of events constituting a reflex action.

For example, let *A* be the skin and *B* a certain group of muscles. If the skin be touched, afferent impulses travel inwards to nerve-centres in the ganglia (*C*), which thereupon send forth efferent impulses to the appropriate muscles. Orderly contractions result, and the worm draws back from the unwelcome irritation. Within the nerve-centre some discriminating action takes place, such that the afferent impulses are not merely " reflected," as the word " reflex " might seem to indicate, but are worked over and are sent as efferent impulses to only those muscles necessary to effect the appropriate action. Little is known of the nature of this discriminating function, but it is certainly unconscious in a vast majority of cases. For a fuller discussion the student is referred to special works on Physiology.

Sensitive System. We have learned (p. 118) that the earthworm possesses the senses of touch, taste, sight, and smell, although it has no special organs for these senses, and hence can hardly be said to possess any proper sensory *system*. We do not know, moreover, whether the so-called "senses" of the earthworm are really states of consciousness as in ourselves, for we do not even know whether earthworms possess any form of consciousness (compare p. 6). When, therefore, we speak of the earthworm as possessing the "sense" of touch or of sight we mean simply that some of the nerves terminating in the skin may be stimulated by mechanical means or by rays of light, without necessarily implying that the worm actually feels or sees as we feel and see.

The sense of *touch* extends over the whole surface of the body. That of *taste* is probably located in the cavity of the mouth and pharnyx; the location of the sense of *smell* is unknown. Darwin's experiments have shown that the earthworm's feeble sense of *sight* is confined to the anterior end of the body. It is probable that the nerves of sight, taste, and smell enter the cerebral ganglia alone, while those of touch run to other ganglia as well.

Systems of Support, Connection, Protection, etc. The structure and mode of life of many animals are such as to require some solid support to the soft parts of the body. Such supporting structures are, for instance, the bones of vertebrata, the hard outer shell of the lobster, or beetle, and the coral which forms the skeleton of a polyp. The earthworm has, however, nothing of the sort, and it is obvious that a hard supporting organ would be not only useless, but even detrimental. The power of creeping and burrowing through the earth depends upon great flexibility and extensibility of the body; and with this the presence of a skeleton would be incompatible.

The *connecting* system consists simply of various tissues and substances by which the different organs are bound firmly together. These can only be seen upon microscopical examination. The most important of them is known as *connective tissue.*

As to *protective* structures, the earthworm is probably one of the most defenceless of animals. Nevertheless there are certain structures which are clearly for this purpose. The *cuticle* which covers the surface is a thin but tough membrane which protects the delicate skin from direct contact with hard objects. It passes into the mouth and anus and lines the entire alimentary canal. In the gizzard, where food is ground up, it is prodigiously thick and tough, and must form a very effective protection for the soft tissues beneath it. The main defence of the animal lies, however, not in any special armor, but in those instincts which lead it to lie hidden in the earth during the day and to venture forth only in the comparative safety of darkness.

CHAPTER VIII.

THE BIOLOGY OF AN ANIMAL (*Continued*).

THE EARTHWORM.

Reproduction. The general nature of the reproductive function has already been fully explained (p. 88) and illustrated in the case of the fern. Unlike the fern, the earthworm multiplies only by *sexual* reproduction or *gamogenesis*. This animal must therefore be compared as to its reproduction with the oöphore, not with the sporophore, of *Pteris*, and there is no alternation of generations (cf. p. 89).

Germ-cells. Here, as in the fern, male and female cells are borne by the same individual, so that the earthworm is bi-sexual or hermaphrodite. The male and female germ-cells are very similar to those of the fern, the former (*spermatozoa*) being very minute and actively mobile, while the latter (*ova* or *eggs*) are much larger and immobile. The ova are produced in special organs, the *ovaries*, and the *spermatozoa* in organs known as *spermaries* or *testes*.

The ripe ovum (Fig. 83, *ov*) is a nearly typical cell, agreeing closely with that of the star-fish (Fig. 20), except that it has no membrane. It is still customary to apply to ova the old terminology, calling the cell-substance *vitellus*, the membrane *vitelline membrane*, the nucleus *germinal vesicle*, and the nucleolus *germinal spot*.

The ripe spermatozoön (Fig. 83, *sp*) is an extremely minute elongated cell or filament tapering towards one end (*tail*) to a point, and having a slight enlargement at the other, the *head*. The latter is really the nucleus of the cell enveloped by a thin layer of protoplasm. The tail of an active spermatozoön is virtually a long cilium (p. 34), which by vigorous lashing drives the whole cell along head-foremost, very much as a tadpole is driven by its tail.

Since the ovaries and spermaries give rise to the germ-cells, they are called the *essential organs* of reproduction. Besides these, *Lumbricus*, like most animals, has *accessory organs* of reproduction which act as reservoirs or carriers of the germs, assist in securing cross-fertilization, and minister to the wants of the young worms.

Essential Reproductive Organs. The *ovaries* are two in number and lie one on either side in the 13th somite attached to the hinder face of the anterior dissepiment (*ov*, Fig. 80). They are about 2^{mm} in length, distinctly pear-shaped, and attached by the broader end (Fig. 82). The narrow extremity contains a single row of ova and is called the *egg-string* (*es*). In this the ova are ripe or nearly so; behind they shade off into those more and more immature till these are lost in a mass of nearly undifferentiated cells (*primitive ova*), constituting the great bulk of the ovary. Each of these, however, is surrounded with still smaller cells constituting its nutrient envelope or *follicle*. As the ova mature the follicles still persist, and they may be detected even in the egg-string. When fully ripe the ovum bursts the follicle and is shed from the end of the egg-string into the body-cavity, where it floats for a time in the coelomic fluid. It is ultimately taken into the oviduct and carried to the exterior.

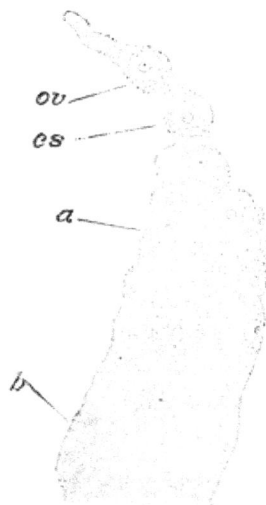

FIG. 82. The ovary, much enlarged; *b*, the basal part, *a*, body of the ovary containing immature ova; *es*, egg-string; *ov*, ripe ovum ready to fall off.

The *spermaries* or *testes* (*t.t*, Fig. 80) are six in number, and are perhaps the most conspicuous of all the internal organs of the earthworm. They are large white bodies arranged in three pairs united into one lobed mass which lies between the 10th and 12th somites, and partly covers the oesophagus. Towards the median line they are connected with a large sac (*s.v*, Fig. 80) underlying the oesophagus. This is the *seminal vesicle*, a reservoir for the storage of the ripe spermatozoa which are produced in the various lobes of the testes. A median partition divides it into a fore part connected with the anterior pair, and a hinder part connected with the middle and posterior pair of testes. These chambers are usually regarded as two separate seminal vesicles, but are often scarcely distinct, as in the figure. The spermatozoa are differentiated in the spermaries from cells forming the internal mass; they are at first rounded like the primitive ova, but eventually become as it were drawn out into the form of a filament. They are at first motionless, but gradually acquire the power of movement and swim swiftly and vigorously about.

Accessory Reproductive Organs. After the ova and spermatozoa are ripe they are finally transported to the outside of the animal by the *oviducts* and *sperm-ducts*. The former (*o.d*, Fig. 80) are two short trumpet-shaped tubes, opening externally by the smaller end upon the 14th somite. The inner end opens freely by a wide ciliated funnel into the cavity of the 13th somite. The oviducts are placed just behind the ovaries, above and a little to the side of the nephridial funnel of the 13th somite. The external opening lies between the two groups of setae on the 14th somite (Fig. 72, *o.d*). By means of the ciliated funnel the floating ova are drawn into the oviduct and cast out upon the outer surface of the body.

The *sperm-ducts* (*vasa deferentia*) (*s.d*, Fig. 80) convey the sperm (including the spermatozoa and the liquid in which they move) to the surface of the 15th somite, where they open by a conspicuous slit surrounded by fleshy lips (Fig. 72, *s.d*). They are very long, slender tubes, open like the oviducts at both ends, and somewhat trumpet-shaped; they run straight forward to the 12th somite, where each branches like a Y. The two branches pass obliquely forwards and inwards to join the seminal vesicle (Fig. 80). The hinder branch joins the posterior vesicular chamber. The other proceeds to the anterior chamber, with which it becomes connected (Fig. 80). Just before joining the vesicle each tube is twisted into a peculiar knot, and at the vesicle is dilated into a funnel-shaped orifice closely attached to the lower side of the vesicles. The sperm-ducts are ciliated, and closely resemble the oviducts in mode of action.

The obvious likeness between the reproductive ducts and nephridia suggests the idea that the former may have been derived from the latter. It is an interesting fact that in many worms nephridia act as ducts for the emission of the reproductive elements, and that *Lumbricus* is exceptional in having special ducts for this purpose. It has been suggested that the earthworm perhaps had originally two pairs of nephridia in each somite, of which one pair has disappeared from each somite, excepting the 14th and 15th, where they still persist as the oviducts and sperm-ducts.

The *seminal receptacles* are accessory organs of reproduction in the shape of small rounded sacs or pouches, open to the outside only, at about the level of the upper row of setæ. They lie between the 9th and 10th, and 10th and 11th somites (*s.r.* Figs. 75 and 80), where their openings may be sought for (Fig. 72). Their function is explained under the head of copulation.

Accessory glands. Besides all the structures so far described there are many glands which play a part in the reproductive functions. The setigerous glands from about the 7th to about the 19th somite (sometimes fewer, sometimes none at all) are often greatly enlarged, and form the glandular prominences mentioned at p. 124. They seem to be used as organs of adhesion during copulation. The clitellum is filled with gland-cells which probably serve in part to secrete a nourishing fluid for the young worms, and in part to provide a tough protecting membrane to cover them.

Copulation. Egg-laying. Inasmuch as each individual earthworm produces both ova and spermatozoa, it might be supposed that copulation, or the sexual union of two different individuals, would not be needful. This, however, is not the case. The ova of one individual are invariably fertilized by the spermatozoa of another individual after a process of copulation and exchange of spermatozoa, as follows: During the night-time, and usually in the spring, the worms leave their burrows and pair, placing themselves so that their heads point in opposite directions and holding firmly together by the enlarged setigerous glands and the thickened lower lateral margins of the clitellum. During this act the seminal receptacles of each worm are filled with spermatozoa from the sperm-ducts of the other, after which the worms separate.

When the worm is ready to lay its eggs the glands of the clitellum become very active, pouring out a thick glairy fluid which soon hardens into a tough membrane and forms a girdle around the body. Besides this a large quantity of a thick jelly-like nutrient fluid is poured out and retained in the space between the girdle and the body of the worm. The girdle is thereupon gradually worked forward toward the head of the worm by contractions of the body. As it passes the 14th somite a number of ova are received from the oviducts, and between the 9th and 11th

10

somites a quantity of spermatozoa are added from the seminal receptacles where they have been stored since the time of copulation when they were obtained from another worm. The girdle is next stripped forwards over the anterior end and is finally thrown completely off. As it passes off its open ends immediately contract tightly together, and the girdle becomes a closed capsule (Fig. 83, *c*) containing both ova and spermatozoa floating in a nutritive fluid or milk. The membrane soon assumes a light yellowish or brown color, becomes hard and tough, and serves to protect the developing embryos. The capsules may be found in May or June

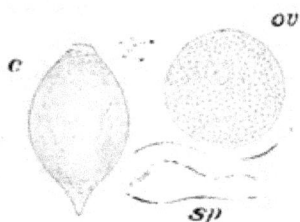

Fig. 83.—Egg capsule enlarged 5 diameters; a few eggs enlarged to the same scale are shown near by; *ov*, an ovum enlarged; *sp*, spermatozoa (these are somewhat too large).

in earth under logs or stones, or especially in heaps of manure. Within the capsules the fertilization and development of the ova take place.

EMBRYOLOGY.

Fertilization. The spermatozoa swim actively about in the fluid which fills the capsule, approach the ova, and attach themselves to the latter by their heads in large numbers. The tails still lash actively to and fro, and the spermatozoa appear to make vigorous efforts to bore their way into the ova. Sooner or later a single spermatozoon forces its head into each ovum, leaving outside its tail, which soon loses the power of movement and finally dies. When inside the ovum the head becomes the centre of a star-shaped structure formed by a radiating arrangement of the vitellus in its neighborhood. This is known as the *male pronucleus.*

Meanwhile the germinal vesicle undergoes great changes, which finally result in the formation of a second star-shaped structure called the *female pronucleus.* Lastly the male and female pronuclei travel towards one another and finally fuse completely together, forming a single body known as the *segmentation-nucleus.*

The whole process just described is known as fertilization or impregnation of the ovum, and the product is called the fertilized ovum. The latter is essentially similar in structure to the origi-

nal ovum, but it has now become a dual being which has resulted
from the fusion of two cells, and it is now capable of development into a new individual. The process of fertilization appears
to be essentially the same among all higher animals, and in a
broader sense to be identical with the sexual process among all
higher and many lower plants. (Compare the fern, p. 98.)

Cleavage of the Fertilized Ovum. Soon after fertilization the
ovum begins the remarkable process of multiplication which has
already been briefly sketched at p. 51, Chap. IV. The segmentation-nucleus divides into two parts, and this is followed by
a division of the vitellus, each half of the original nucleus

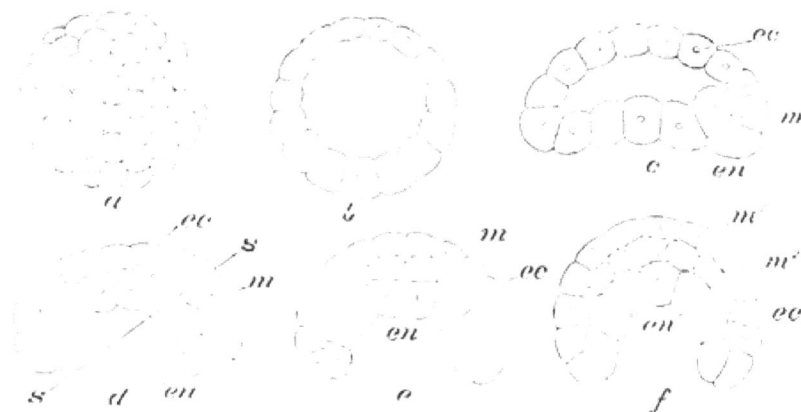

FIG. 81.—Diagram of the early stages in the development of the earthworm: *a,* the blastula in surface view; *b,* the same in section showing the large segmentation-cavity; *c,*
beginning of the invagination to form the gastrula; *d,* the gastrula in section; *e,* section
of the gastrula along the straight line *ss* shown in *d* ; *f,* older gastrula showing the
growth of the mesoblast; *ec,* ectoblast; *en,* entoblast; *m, m',* primary mesoblastic cells.
m², mesoblast.

becoming the nucleus of one of the halves of the vitellus; that
is, the original cell divides into two smaller but similar cells (see
Fig. 21). These divide in turn into four, and these successively
into eight, sixteen, etc., until the embryo consists of a large number of cells. At the division of the four cells into eight, four are
smaller and four larger, so that an upper and a lower pole may
be distinguished in the embryo. These poles can be recognized
throughout the remainder of this period of development, the cells
of the upper half of the embryo remaining smaller and less granular than those of the lower half.

Almost from the first the cells arrange themselves so as to

surround a central cavity known as the *segmentation-cavity*. This cavity increases in size in later stages, so that the embryo finally appears as a hollow sphere surrounded by a wall consisting of a single layer of cells. This stage is known as the *blastula* (or *blastosphere*) (*a, b*, Fig. 84).

Formation of the Germ-Layers. The blastula still consists of undifferentiated cells, and has no organs. From this time forwards the whole course of development is a process of differentiation, both of the cells and of the organs into which they soon arrange themselves. The first step in this process is a flattening of the embryo at the lower pole—i.e., the half consisting of larger cells (Fig. 84, *c*). The large cells are then folded into the segmentation-cavity so as to form a pouch opening to the exterior; at the same time the embryo becomes somewhat elongated.

This process is known as *invagination*, and at its completion the embryo is called the *gastrula*. The infolded pouch (called the *archenteron*) is the future alimentary canal; its opening (now known as the *blastopore*) will become the mouth; and the layer of small cells over the outside will form the skin or outer layer of the body-wall.

The embryo very soon begins to swallow, through the blastopore, the milk-like fluid in which it floats, and to digest it within the cavity of the archenteron.

It is obvious that the embryo already shows a distinct differentiation of parts which perform unlike functions. In fact we may regard the gastrula as composed of two tissues still nearly similar in structure but very unlike in function. One of these consists of the layer of cells which forms the outer covering: this tissue is known as the *ectoblast* (*ec*, Fig. 84). The second tissue is the layer of cells forming the wall of the archenteron: it is called the *entoblast* (*en*). The ectoblast and entoblast together are known as *germ-layers*.

Meanwhile changes are taking place which result in the formation of a third germ-layer lying in the segmentation-cavity between the ectoblast and entoblast and therefore called the *mesoblast* (*m*, Figs. 84, 85). As the entoblast is invaginated two large cells separate from it and pass into the segmentation-cavity (Fig. 84, *c*). These are known as *primary mesoblastic cells* because they afterwards give rise to the mesoblast. Each of these cells rapidly multiplies to form an elongated mass of cells extending at each side of the archenteron. After each division the primary mesoblastic cells increase in size so that up to a late stage in

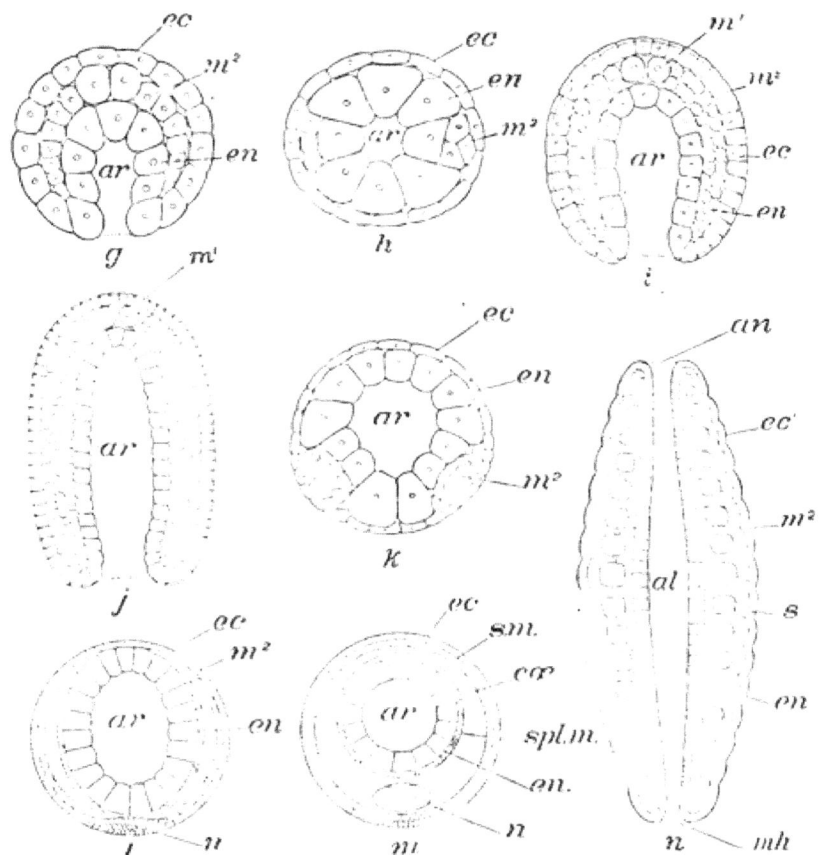

FIG. 85.—Diagrams of later embryonic stages: *g*, embryo in long section, showing the germ-layers, archenteron, and blastopore; *h*, the same in cross-section, showing the mesoblastic bands; *i*, later stage; *j*, still later stage in longitudinal section, showing the appearance of the cavities of the somites; *k*, the same in cross-section; *n*, diagram of a young worm in longitudinal section after the formation of the stomodaeum, proctodaeum, and anus; *l*, the same in cross-section, showing the beginning of the nervous system; *m*, cross-section of later stage with the nervous system completely established. *al*, alimentary canal; *ar*, archenteron; *an*, anus; *cœ*, coelom; *ec*, ectoblast; *en*, entoblast; *m'*, primary mesoblastic cells; *m²*, mesoblast; *mh*, mouth; *n*, nervous system; *s*, cavity of somite; *s.m.*, somatic layer of the mesoblast, which with the ectoblast forms the somatopleure; *spl.m*, splanchnic layer of the mesoblast, which with the entoblast forms the splanchnopleure.

development they may be distinguished from the cells to which they give rise. The two masses of mesoblastic cells gradually increase in size and finally fill the segmentation-cavity.

In some species of *Lumbricus* the mesoblast is in part formed by a multiplication of the ectoblastic cells at their inner ends. In other species no large primary mesoblastic cells can be distinguished.

The appearance of the ectoblast, entoblast, and mesoblast is known as the formation (or differentiation) of the GERM-LAYERS, and is one of the most important and significant processes in the whole course of development. Germ-layers like those of *Lumbricus*, and called by the same names, are found in the embryos of all higher animals; and it will hereafter appear that this fact has a profound meaning.

Development of the Organs. The embryo gradually increases in size and at the same time elongates. As it lengthens, the blastopore (in this case the *mouth*) remains at one end, which is therefore to be regarded as anterior, and the elongation is backwards. The two primary mesoblastic cells are carried backwards, and always remain at the extreme posterior end (*i, j*, Fig. 85). The cells of all three germ-layers continually increase in number by division, new matter and energy being supplied from the food which is swallowed by the embryo in such quantities as to swell up the body like a bladder. The archenteron enlarges until it comes into contact with the ectoblast and the segmentation-cavity is obliterated. The mesoblast is in the form of two bands lying on either side of the archenteron, and extending forwards from the primary mesoblastic cells.

This is clearly seen in a cross-section of the embryo, as in Fig. 85, *h, k*. The mesoblastic bands are at first solid (*h*), but after a time a series of paired cavities appears in them (*j*), continually increasing in number by the formation of new cavities near the hinder end of the bands as they increase in length. A cross-section passing through one pair of these cavities is shown in *k, l*, Fig. 85. As the bands lengthen they also extend upwards and downwards (*l*, Fig. 85), until finally they meet above and below the archenteron (*m*). The cavities at the same time continue to increase in size, and finally meet above and below the archenteron, which thus becomes surrounded by the body-cavity or cœlom. The cavities are separated by the double partition-walls of mesoblast. These partitions are the dissepiments, and the cavities themselves constitute the cœlom. The outer mesoblastic wall of each cavity is known as the *somatic layer* (*s.m.*): it unites with the ectoblast to constitute the body wall (*somato-*

pleura). The inner wall, or *splanchnic layer* (*spl.m*) unites with the entoblast to constitute the wall of the alimentary canal (*splanchnopleure*). An ingrowth of ectoblast (*stomodaeum*) takes place into the blastopore to form the pharynx, and a similar ingrowth at the opposite extremity (*proctodaeum*) unites with the blind end of the archenteron to form the anus and terminal part of the intestine.

As to its origin, therefore, the alimentary canal consists of three portions, viz., (1) the archenteron, consisting of the original

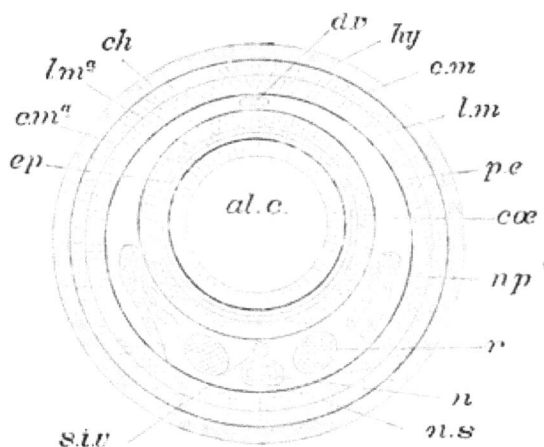

Fig. 86.—Diagram of a cross-section of *Lumbricus*, showing the relation of the various organs, etc., to the germ-layers. Ectoblastic structures shaded with fine parallel lines, entoblastic with coarser parallel lines, mesoblastic with cross-lines; *al.c*, alimentary canals; *ch*, chloragogue layer; *coe*, coelom; *c.m*, circular muscles of body-wall; *c.m.a*, circular muscles of alimentary wall; *ep*, lining epithelium of alimentary canal; *d.v*, dorsal vessel; *hy*, hypodermis or skin; *l.m*, longitudinal muscles of body-wall; *l.m.a*, longitudinal muscles of alimentary wall; *n*, central part of nerve-cord; *np*, nephridium; *n.s*, sheath of nerve cord; *p.e*, peritoneal epithelium; *r*, reproductive organs; *s.i.v*, sub-intestinal vessel.

entoblast; (2) the stomodaeum or pharyngeal region, lined by ectoblast; and (3) the proctodaeum or hindmost part, also lined by ectoblast. These three parts are called the *fore-gut* (stomodaeum), *mid-gut* or mesenteron (archenteron), and *hind-gut* (proctodaeum), and it is a remarkable fact that these same parts can be distinguished in all higher animals, not excepting man.

The body now becomes jointed by the appearance of transverse folds opposite the dissepiments, and the metamerism of the body becomes evident on the exterior. The young worm has thus reached a stage (*n*, Fig. 85) where its resemblance to the

adult is obvious.   It has an elongated, jointed body, traversed
by the alimentary canal, which opens in front by the mouth and
behind by the anus.   The metamerism is expressed externally
by the jointed appearance, internally by the presence of paired
cavities (cœlom) separated by dissepiments.   Both the body-wall
and the alimentary wall consist of two layers: the former of
ectoblast without and somatic mesoblast within: the latter of
splanchnic mesoblast without (i.e., towards the body-cavity), and
either entoblast or ectoblast within, according as we consider the
mid-gut on the one hand, or the fore- and hind-gut on the other.
This is shown in Fig. 86, which represents a cross-section of the
embryo through the mid-gut.   If this be clearly borne in mind
the development of all the other organs is easy to understand,
since they are formed as thickenings, outgrowths, etc., of the
parts already existing.   For instance, the blood-vessels make
their appearance everywhere throughout the mesoblast, and the
nephridia are formed as outgrowths from the dissepiments.   The
reproductive organs are at first mere thickenings on the somatic
layer of the mesoblast, afterwards separating more or less from it
so as to lie in the cavity of the cœlom.   The nervous system is
produced by thickenings and ingrowths from the ectoblast.   The
origin of the different parts is shown in the following scheme:—

THE GERM-LAYERS AND THEIR DERIVATIVES.

Ectoblast.	Outer skin (Hypodermis and Cuticle). Nerves and Ganglia Lining membrane of pharynx (fore-gut). Lining membrane of anus and hinder part of intestine (hind-gut).
Mesoblast.	Muscles. Blood-vessels. Reproductive organs. Nephridia. Outer layers of alimentary canal
Entoblast.	Lining membrane of greater part of the alimentary canal—mid-gut.

The above statements as to the origin of the various organs
acquire great interest in view of the fact that they are essentially true of all animals above the earthworm, as well as of
many below it—of all, in a word, in which the three germ-layers
are developed, i.e., all those above the *Cœlenterata* or polyps,
jelly-fishes, hydroids, sponges, etc.   In man, as in the earthworm

and all intermediate forms, the ectoblast gives rise to the outer skin (epidermis), the brain and nerves, fore- and hind-gut; the entoblast gives rise to the lining membrane of the stomach, intestines, and other parts pertaining to the mid-gut; while the somatic and splanchnic layers of the mesoblast give rise to the muscles, kidneys, reproductive organs, heart, blood-vessels, etc. It is now generally held that the germ-layers throughout the animal kingdom (with the partial exception of the *Calenterata* already mentioned) are essentially identical in origin and fate. This view is known as the *Germ-layer Theory.* It is one of the most significant and important generalizations which the study of Embryology has brought to light, since it recognizes a structural identity of the most fundamental kind among all the higher animals.

Sooner or later the young earthworm bursts through the walls of the capsule and makes its entry into the world. When first hatched it is about an inch long and has no clitellum.

It is a curious fact that in certain species of *Lumbricus* the young worms are almost always hatched as *twins*, two individuals being derived from a single egg by a process which is described by Kleinenberg in the *Quarterly Journal of Microscopical Science,* Vol. XIX., 1879. It often happens that the twins are permanently united by a band of tissue, as in the case of the well-known Siamese twins.

We have now traced roughly the evolution of a complex many-celled animal from a simple one-celled germ. It is important to notice at this point a few general principles which are true of higher animals in general.

1. The embryological history is a true process of development, — not a mere growth or unfolding of a pre-existing rudiment, as the leaf is unfolded from the bud. Neither the ovum nor any of the earlier stages of development bears the slightest resemblance to an earthworm. The embryo must undergo a transformation of structure as well as an increase of size.

2. It is a progress from a one-celled to a many-celled condition.

3. It is a progress from relative simplicity to relative complexity. The ovum is certainly vastly more complex than it appears to the eye, but no one can doubt that the full grown worm is more complex still.

4. It is a progress from a slightly differentiated to a highly

differentiated condition. The life of the ovum is confined to a single cell. The blastula is composed of a number of nearly similar cells, which in the gastrula become differentiated into two distinct tissues. In later stages the cells become differentiated into many different tissues, which in turn build up different organs performing unlike functions.

5. Lastly, it is a cycle, beginning with the ovum, and after all the complicated changes of development resulting in the production of new ova, which repeat the process and give rise to a new generation. All other cells in the body (some of the spermatozoa of course excepted) must sooner or later die. The ovum alone persists as the starting-point to which the cycle of life continually returns (cf. p. 110).

CHAPTER IX.

THE BIOLOGY OF AN ANIMAL (*Continued*).

THE EARTHWORM.

MICROSCOPIC STRUCTURE OR HISTOLOGY.

WE have followed the development of the one-celled germ through a stage, the *blastula*, in which it consists of a mass of nearly similar cells out of which the various tissues of the adult eventually arise. The first step in this direction is the differentiation of the *germ-layers* or three primitive tissues (p. 148). As the embryo develops, the cells of these three tissues become *differentiated in structure* to fit them for different duties in the physiological division of labor. And when this process of differentiation is accomplished and the adult state is reached we find six well-marked varieties of tissue, as follows:—

PRINCIPAL TISSUES OF *Lumbricus.*

I. **Epithelial.** Layers of cells covering free surfaces.
> (*a*) *Pavement Epithelium.* Cells thin and flat, arranged like the stones of a pavement.
> (*b*) *Columnar Epithelium.* Cells elongated, standing side by side, palisade-like.
> (*c*) *Ciliated Epithelium.* Columnar or cuboid, and bearing cilia.

II. **Muscular.** Cells contractile and elongated to form *fibres*. Often arranged in parallel masses or *bundles.*

III. **Nervous.** Cells pear-shaped or irregular, with large nuclei; having processes prolonged into slender cords or fibres, bundles of which constitute the *nerves.*

IV. **Germinal.** Including the germ cells. At first in the form of epithelial cells covering the coelomic surface, but afterwards differentiated into ova and spermatozoa.

V. **Blood.** Isolated cells or corpuscles floating in a fluid intercellular substance, the *plasma.*

VI. **Connective.** Cells of different shapes, often branched but sometimes rounded, separated from one another by more or less of lifeless intercellular substance in the form of threads or homogeneous material.

These six kinds of tissue constitute the main bulk of the earthworm, as of higher animals generally; but there are in addition other tissues which will be treated of hereafter.

Arrangement of the Tissues. The simplest and most direct mode of discovering the arrangement of the tissues is by the microscopical study of thin transverse or longitudinal sections (prac-

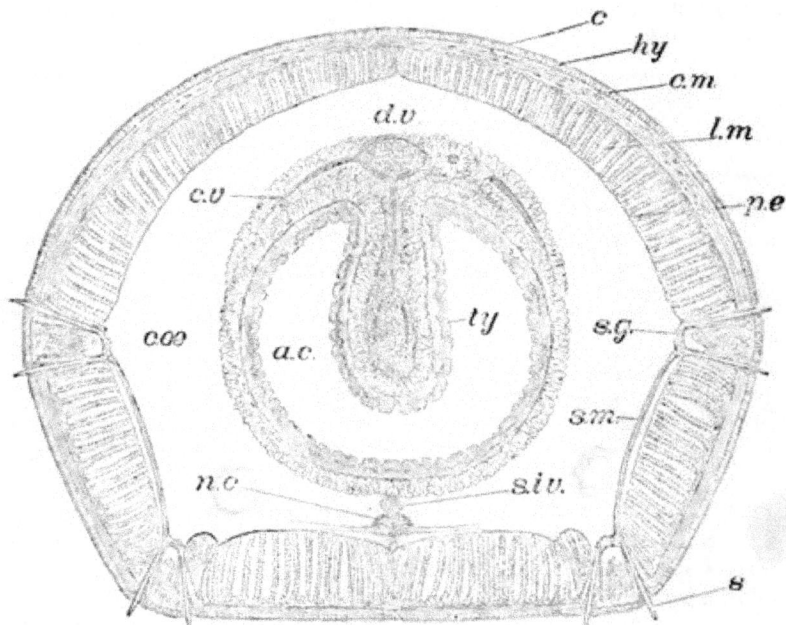

FIG. 87.—Transverse section of the body behind the clitellum. *a.c*, cavity of the alimentary canal; *c*, cuticle; *cœ*, cœlom; *c.m*, circular muscles; *c.v*, circular vessel; *d.v*, dorsal vessel; *hy*, hypodermis; *l.m*, longitudinal muscles; *n.c*, ventral nerve-chain; *p.e*, peritoneal epithelium; *s*, seta; *s.g*, setigerous gland; *s.i.v*, subintestinal vessel; *s.m*, muscle connecting the two groups of setæ on the same side; *ty*, typhlosole.

tical study, p. 171). A transverse section taken through the region of the stomach-intestine is represented in Fig. 87. Its composition is as follows :—

A. Body-wall.

This consists of five layers, viz. (beginning with the outside).—

1. *Cuticle* (*c*). A very thin transparent membrane, not composed of cells and perforated by fine pores. It is a product or secretion of the —

2. *Hypodermis* (*hy*) (epidermis or skin). A layer of colum-

nar epithelium, composed of several kinds of elongated cells, set
vertically to the surface of the body. Some of these, known as
gland-cells, have the power of producing within their substance
a glairy fluid (mucus), which exudes to the exterior through the
pores in the cuticle. The mucus often accumulates within these
cells to such an extent as to swell them up, giving them an oval
form and clear appearance. Many of the cells are branched at
their inner ends.

3. *Circular Muscles (c.m).* A layer of parallel muscle-fibres
running around the body. On the upper side they are intermingled with connective-tissue cells containing a granular brown-
ish substance (pigment) which gives the dorsal aspect its darker tint.

4. *Longitudinal Muscles (l.m).* A layer of muscle-fibres running lengthwise of the body. They are arranged in complicated
bundles, which in cross-sections have a feathery appearance. In
longitudinal sections they appear as a simple layer, and resemble
the circular fibres as seen in the cross-section.

The circular muscles are arranged in somewhat similar bundles, as may be seen in longitudinal sections.

5. *Peritoneal Epithelium (p.e).* A very thin and delicate
layer of flattened cells attached by their edges and covering the
inner face of the muscular layers. This is only visible in very
good sections. It extends over the whole surface of the cœlom,
covering also the organs which the cœlom contains.

The hypodermis, and therefore also the cuticle to which it
gives rise, is derived from the ectoblast. The other layers (3, 4,
5) arise from the somatic layer of the mesoblast.

B. Alimentary Canal.

The wall of this tube appears in cross-section as a ring surrounded by the cœlom. The typhlosole (*ty*) is seen to be a deep
infolding of its upper portion. The wall is composed of six
layers as follows, starting from the alimentary cavity:—

1. *Cuticle (c, Fig. 88).* A transparent structureless membrane like the outer cuticle, with which it is continuous through
the mouth and anus. In the gizzard it is enormously tough and
thick, and is roughened by minute points.

2. *Lining Epithelium (l.p).* A layer of closely packed, narrow, columnar cells with oval nuclei.

3. *Vascular Layer (v.l).* Numerous minute blood-vessels.

4. *Circular Muscles (c.m)*. A thin layer of muscle-fibres running around the gut.

5. *Longitudinal Muscles (l.m)*. A thin layer of muscle-fibres running along the gut.

6. *Chloragogue Layer (ch)*. Composed of large polyhedral or rounded cells containing yellowish-green granules. The cells fill the hollow of the typhlosole, and cover the surface of the dorsal and lateral blood-vessels.

7. *Peritoneal Epithelium* (not shown in the figure).

The same general arrangement exists in all parts of the alimentary canal, but is sometimes greatly modified. For instance, in the gizzard and pharynx the muscular layers are enormously developed and the cuticle is thick. In a part of the gizzard the chloragogue-layer is nearly or quite absent and the typhlosole disappears. A fuller description of these modifications will be found in Brooks's *Handbook of Invertebrate Zoölogy*, and a complete account in Claparède, *Zeitschrift für wissenschaftliche Zoologie*, Vol. XIX., 1869.

The lining epithelium and the cuticle are derived from the entoblast. The remaining layers arise by differentiation of the splanchnic layer of mesoblast.

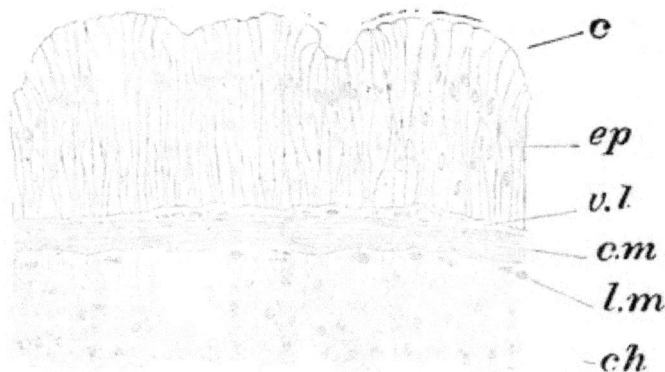

FIG. 88.—Highly magnified cross-section through the wall of the alimentary canal. *c*, cuticle on the inside; *ch*, chloragogue layer; *c.m*, circular muscles; *ep*, lining epithelium; *l.m*, longitudinal muscles; *v.l*, vascular layer.

Blood-vessels appear in the section as rounded or irregular cavities bounded by thin walls. They consist of a delicate lining epithelium covered by a thin layer of muscle-fibres. In the walls of the stomach-intestine the vessels are often completely invested by chloragogue-cells, which radiate from them with great regularity (Fig. 87). The finer branches have no muscular layer, consisting of the epithelium alone.

Dissepiments These often appear in cross or longitudinal sections They consist chiefly of muscle-fibres irregularly disposed, intermingled with connective-tissue cells and fibres, and covered on both sides with the peritoneal epithelium.

Nervous System A cross-section of a ganglion (Fig. 89) shows it to be composed of two distinct parts, viz., (1) the gan-

FIG. 89.—Highly magnified cross-section of a ventral ganglion; *g f,* giant-fibres; *l.n,* lateral nerve; *n c,* nerve cells; *s,* muscular sheath of the ganglion; *s.v,* sub-neural vessel; *s.n v,* supra-neural vessel.

glion proper on the inside, and (2) a sheath which envelops it. The sheath (*s,* Fig. 89) consists of two layers, viz.:—

1. *Peritoneal Epithelium.* On the outside.

2. *Muscular Layer,* or sheath, a thick layer of irregularly arranged muscle-fibres intermingled with connective-tissue. Imbedded in it are the sub-neural blood-vessel on the lower side and the supra-neural blood-vessels on each side above. In the middle line are three rounded spaces (*g. f,* Fig. 89), which are the cross sections of three hollow fibres of unknown nature running along the entire length of the ventral nerve-chain. They are called "giant-fibres," and possibly serve to support the soft parts of the nerve-cord.

The *Ganglion* proper is distinctly bilobed, and consists of two portions, viz.:—

1. *Nerve-cells (n.c).* Numerous pear-shaped nerve-cells near the surface, with their narrow ends turned towards the centre,

into which each sends a single branch or nerve-fibre. They are confined to the ventral and lateral parts of the ganglion.

2. *Fibrous Portion.* This occupies the central part. It consists of a close and complicated network of nerve-fibres intermingled with connective-tissue. Some of these fibres communicate with branches of the nerve-cells, as stated above; others run out into the lateral nerves, while still others run along the commissures to connect with fibres from other ganglia.

A section through the brain or supra-œsophageal ganglia shows the same essential structure, but the muscular layer of the sheath is very thin and the nerve-cells are on the front and upper sides of the ganglia, instead of on the lower.

Sections through the ventral commissures are similar to those through the ganglia, but the central portion (i.e., that within the sheath) is smaller, is divided into two distinct parts, and the nerve-cells are less abundant.

Sections through the nerves show them to consist only of parallel fibres, surrounded by a sheath which gradually fades away as the nerves grow smaller, and finally disappears, the muscular layer first disappearing, and then the epithelial covering.

Other Organs. The structure of the ovaries has already been described. That of the spermaries varies according to the state of development. When young they consist of closely-packed, rounded cells, each of which gives rise to a number of spermatozoa, which are at first attached by their heads in bundles, the tails being parallel. These bundles are easily seen in sections.

The Clitellum is produced by an enormous thickening of the hypodermis, caused especially by a great development of the gland-cells. Three forms of these may be distinguished, which probably produce different secretions. The tissue is permeated by numerous minute blood-vessels which ramify between the cells.

With this brief sketch of the histological structure of the earthworm we conclude our morphological study of the animal. Those who desire fuller information on the histology will find an exhaustive treatment of it in the work of Claparède, already cited at p. 158.

CHAPTER X.

THE BIOLOGY OF AN ANIMAL (*Continued*).

THE EARTHWORM.

GENERAL PHYSIOLOGY.

IN the preceding pages frequent mention has been made of special physiological phenomena in connection with the detailed descriptions of the primary functions and systems. It now remains to consider the more general problems of the life of the animal, and especially its relations to the environment, and the transformations of matter and energy which it effects.

Like the fern, the earthworm is a small mass of (mainly) living matter occupying a definite position in space and time. It is found in a certain environment for which it is very perfectly fitted, and it is descended from a long series of ancestors in which this adaptation has constantly been maintained by survival of the fittest (compare p. 104). Harmony with the environment appears both in the structure and in the habits of the animal. For instance, the delicate integument gives to the body the flexibility demanded by the peculiar mode of locomotion; it affords at the same time a highly favorable respiratory surface—a matter of no small importance to the worm in its badly-ventilated burrow; and yet this delicate integument does not lead to desiccation, because the animal lives always in contact with moist earth. The alimentary canal, long and complicated, is most perfectly fitted for working over and extracting nutriment from the earthy diet. A striking instance of adaptation is shown in the care which is taken to insure the welfare of the embryo worms. Minute, delicate, and helpless as they are, they develop in safety inside the tough, leathery capsule (p. 146), floating in a milk-like liquid which is at once their cradle and their food.

The worm is no less perfectly adapted to its environment in respect to habit. Its nocturnal habits protect it from heat, desiccation, from birds and other enemies; and even at night it keeps

11

on the alert, retreating at the slightest alarm. In winter or in seasons of drought it burrows deep into the earth.

Income. As in the case of other living things, the life of *Lumbricus* is carried on by a continual interchange of matter and energy with the environment; but the character of the exchange differs widely from that of the fern, chiefly for the reason that the animal cannot directly utilize the kinetic energy of light as a source of vital power. Heat is directly absorbed, it is true, but this is not available except as a favorable *condition* for vital action. The sole source of energy must therefore be ready-made food (carbohydrates, proteids, etc.), which has been manufactured by plants from crude materials such as carbon dioxide and water. Moreover, *proteid matter* is an essential constituent of this food. We do not know why the earthworm must have proteids, but it is a fact; and *Lumbricus* is a type of all animals in this respect. The worm also requires free oxygen, salts, and water, but in this it agrees with all living things. Probably it lives best upon a mixed diet containing proteids, carbohydrates, fats, water, salts, and free oxygen. It is difficult to determine the exact income of *Lumbricus*, but it may be set down approximately as follows :—

INCOME OF *LUMBRICUS.*

MATTER.	WHENCE DERIVED.
1. *Proteids.*	From vegetal or animal matters taken in through the mouth.
2. *Fats.*	From vegetal or animal matters taken in through the mouth.
3. *Carbohydrates.*	From vegetal or animal matters taken in through the mouth.
4. *Water.*	Taken in through the mouth, or perhaps to some extent absorbed through the body-walls.
5. *Free oxygen.*	Absorbed directly from the atmosphere or ground-air by diffusion through the body-walls.
6. *Salts.*	Various inorganic salts taken along with other food-stuffs.
ENERGY.	
Potential.	In the food.
Kinetic.	Heat directly absorbed.

Digestion and Absorption. *Lumbricus* takes daily into its alimentary canal a certain amount of necessary food-stuffs, but these are not really inside the body so long as they remain in the alimentary canal; for this is shown by its development to be only a part of the outer surface folded in to afford a safe receptacle within which the food may be worked over. Before the food can be actually taken into the body, or *absorbed*, it must undergo certain chemical changes collectively called *digestion*. The most essential part of this process consists in rendering non-diffusible substances diffusible, in order that they may pass through the walls of the alimentary canal into the blood. Proteids, for example, have been shown to be non-diffusible (Chap. III.). In digestion they are changed by the fluids of the alimentary canal into *peptones* — substances closely like proteids, but readily diffusible. In like manner the non-diffusible starch is changed into diffusible sugar (glucose, $C_6H_{12}O_6$, and becomes capable of absorption. It is highly probable that all carbohydrates are thus turned into glucose or grape-sugar. The fats are probably converted in part into soluble and diffusible soaps which are readily absorbed; and in part emulsified and directly passed into the cells of the alimentary tract in a finely divided state, but nothing is known of this save by analogy with higher animals. In every case, however, digestion takes place *outside the body*, and is only preliminary to the real entrance of food into the living matter.

Metabolism. After absorption the incoming matters are distributed by the circulation to the ultimate living units or cells, and are finally taken up by them and built into their substance. There is reason to believe that each cell takes from the common carrier, the blood, only such materials as it needs, leading a somewhat independent life as to its own nutrition. It co-operates with other cells under the direction of the nervous system (co-ordinating mechanisms), but to a great degree is independent in its choice of food — just as a soldier in a well-fed army obeys orders for the common good, but takes what he pleases from the daily ration supplied to all.

What takes place within the cell upon the entrance of the food is almost wholly unknown, but somehow the food-matters, rich in potential energy, are built up into the living substance probably by a series of constructive processes culminating in protoplasm. Alongside these constructive processes a continual destructive action goes on ; for living matter is decomposed and energy set free in every vital action, and vitality or life is a continuous process. Like the syntheses, the decompositions are probably not single actions, but form a long series. Foster has illustrated the metabolic phenomena by a double flight of steps, at the top of which are always free oxygen and living protoplasm as it were aflame. Foods enter the body, pass up into the living mass of protoplasm, are built into its substance and infected with its properties ; sooner or later they are oxidized, and having set free their energy are cast down by steps on the other side and ultimately excreted. This, however, is only an attempt to picture one of the most wonderful and complicated of all biological phenomena. Of the real process we know absolutely nothing save through its results.

It is clear that no metabolic system can be recognized apart from any other. Were any structures present especially devoted to the temporary storage of reserve supplies (such as the fatty tissue of many animals) they might be regarded as such a system. Possibly the chloragogue-cells have such a function, but this is a mere conjecture.

OUTGO OF LUMBRICUS.

MATTER.	MANNER OF EXIT.
Carbon dioxide (CO_2).	Mainly by diffusion through the skin.
Water (H_2O).	Through the skin, through the organs of excretion, and in the faeces.
Urea [(NH_2)$_2CO$], etc.	Through the excretory organs, and possibly in the faeces.
Salts.	Dissolved in the water.
Proteids, etc.	In the substance of the germ-cells, the egg-capsules, and the contained nutrient fluids.
ENERGY.	
Potential.	A small amount still remaining in urea, in the germ-cells, etc.
Kinetic.	Mechanical work. Heat.

Outgo. Both energy and waste matters leave the cells, and finally leave the body—the former as heat, mechanical work done, or energy unused (in urea, etc.); the latter as excretions, which diffuse freely outwards through the skin, nephridial surfaces, alimentary canal, etc. This is shown in the table on p. 164.

Of the daily outgo the water, carbon dioxide, and salts are devoid of energy, but the urea contains a small amount which is a sheer loss to the animal. Were the earthworm a perfect machine it could use this residue of energy by decomposing the urea into simpler compounds [viz., ammonia (NH_3), carbon dioxide (CO_2), and water (H_2O)]; but it lacks this power, though there are certain organisms (Bacteria) which are able to utilize the last traces of energy in urea. To the daily outgo must be added the occasional loss, both of matter and of energy, suffered in giving rise to ova and spermatozoa, and in providing a certain amount of food and protection for the coming generation.

Comparison of the Fern and Earthworm. To the superficial observer the fern and earthworm seem to have little or nothing in common, except that both are what we call alive. But whoever has studied the preceding pages must have perceived beneath manifold differences of detail a fundamental likeness between the plant and animal, not only in the substantial identity of the living matter in the two, but also in the construction of their bodies and in the processes by which they come into existence. Each arises from a single cell which is the result of the union of two differently constituted cells, male and female. In both, the primary cell multiplies and forms a mass of cells, at first nearly similar, but afterwards differentiated in various directions to enable them to perform different functions, i.e., to effect a physiological division of labor. In both, the tissues thus provided are associated more or less closely into distinct organs and systems, among which the various operations of the body are distributed. And in both the ultimate goal of individual existence is the production of germ-cells which form the starting-point of a new and similar cycle.

This fundamental likeness extends also to most of the actions (physiology) of the two organisms. Both possess the power of adapting themselves to the environments in which they live. Both take in various forms of matter and energy from the environment, build them up into their own living substance, and finally break down this substance more or less completely into simpler compounds by processes of internal combustion, setting free by this action the energy which maintains their vital activity. And, sooner or later, both give back to the environment the matter and energy which they have taken from it. In other words, both effect an exchange of matter and of energy with the environment.

Nevertheless the plant and the animal do differ. They differ widely in form, and the plant is fixed and relatively rigid while the animal is flexible and mobile. The body of the plant is solid; that of the animal contains numerous cavities. The plant absorbs food directly through the external surface; the animal partly through the external and partly through an internal (alimentary) surface. The plant absorbs simple chemical compounds from the air and earth, and kinetic energy from sunlight; the animal absorbs, for the most part, complex chemical compounds and makes no nutritive use of the sun's kinetic energy. By virtue of this energy the plant is able to manufacture starch from simple compounds, carbon dioxide and water; the animal lacks this power. The plant can build up proteids from the starch and other compounds of its food; the animal absolutely requires proteids in its food. And by manufacturing proteids within its living substance, the plant is relieved of the necessity of carrying on a process of external digestion in order to render them diffusible for entrance into the body.

And yet, great as these differences appear to be at first sight, all of them, with a single exception, fade away upon closer examination. This exception is the *mode of nutrition*. Plants and animals differ in form because their mode of life differs; but a wider study of biology reveals the existence of innumerable animals which have a close superficial resemblance to plants (corals, sponges, hydroids, etc.), and of many plants which resemble animals, not only in form, but also in possessing the power of active locomotion. The plant has no use for a stomach because it is bathed over its whole surface by food-materials which make their entrance freely by simple diffusion. Moreover, the stomach of the worm, as shown by its development, is really a part of the general outer surface which is folded into the body; and the animal, like the plant, therefore, absorbs nutriment over its whole surface —oxygen through the general outer surface, other food-matters through the infolded alimentary surface.

And in like manner it is easy to show that not one of the dif-

ferences between the plant and animal is fundamentally important
save the *mode of nutrition*.   The worm must have complex food
including proteid matter.   The fern is able to manufacture this
complex food out of simple compounds.   In terms of energy,
the worm requires food rich in potential energy; the fern can
manufacture food from matters poor in energy, building them
up by a chemical synthesis into substances energized by the sun's
rays.

Hence it appears, broadly speaking, that the fern is constructive, and stores up energy; the earthworm is destructive, and
dissipates energy.   And this becomes of immense importance in
view of the fact that the fern is typical in this respect of all green
plants, as the earthworm is typical of all animals.

It will hereafter appear that even this difference, great as it
is, is partly bridged over by colorless plants like yeast, moulds,
bacteria, etc., which have no chlorophyll, are therefore unable to
use the energy of light, and hence must have complex energized
food.   But these organisms do not, like animals, require proteid
matter, being able to extract all needful energy from the simpler
fats, carbohydrates, and salts.   When we consider that the distinctive peculiarities of animals can thus be reduced to the sole
characteristic of dependence on proteid food, we cannot doubt
that the differences between plants are of immeasurably less
importance than their fundamental likeness.   While the subdivision of biology into botany and zoölogy is indispensable for
the purposes of practical study, it is, after all, adopted mainly for
the sake of convenience, and is not marked by any great natural
boundary among living things.

It has been the object of the foregoing chapters to give the
student a general conception of organisms, whether vegetal or
animal; of their structure, growth, and mode of action; of their
position in the world of matter and energy, and of their relations
to lifeless things.   With this preliminary knowledge as a basis,
the student is prepared to take up the progressive study of other
organisms, selected as convenient types or examples to illustrate
the more important outlines of General Biology.   It is convenient to begin with low and simple forms and work thence gradually upwards; and it is further desirable to do so because there
is reason to believe that this course corresponds with the path of

actual evolution. Before entering upon this study, however, it is desirable to take a preliminary survey of the different kinds of living things by a brief consideration of the outlines of zoölogical and botanical classification.

PRACTICAL STUDY OF *LUMBRICUS*.

A. General External Characters.

Observe in well-preserved specimens the general form, the somites (count), clitellum, anterior and posterior ends, mouth and anus, right and left sides, dorsal and ventral aspects, number and arrangement of the setæ. Compare the prostomium with the first somite, and the different somites with one another.

B. Openings.

With a hand-lens observe in a large specimen the openings of—

    a. Sperm-ducts (15th somite).

    b. Oviducts (14th somite).

    c. Seminal receptacles (9–10, 10–11 somites) (difficult).

    d. Pores into the cœlom—one in the middle line of the dorsal aspect of each somite.

    e. (? Nephridia, very difficult to see.)

C. Living Earthworms.

1. Place large, active worms on a moist, rough surface (a wet board will answer), and study carefully the mode of locomotion. The direction of crawling is normally forwards, but often backwards if the anterior extremity be cautiously touched.

2. Note the quick reflex actions called forth by stimuli (taps, etc.).

3. Place a worm on its back and note the results.

4. Place a worm under a bell-jar with a small open dish of chloroform. Note—

    a. Speedy loss of sensibility to stimuli.

    b. Afterwards, pronounced *rigor mortis* (p. 46).

    c. Place a chloroformed worm (not in *rigor*) in water and gradually heat to about 50° C. Note the *rigor caloris*. Compare with *rigor mortis.*

5. Note in a fresh worm the pulsations of the dorsal vessel seen through the skin.

6. Place worms in a large, covered, shallow dish, with bits of

leaves, meat, moistened paper, fat, etc. At night the worms may be seen in the act of eating.

7. If practicable, study by night the habits of worms in their natural homes, with the aid of a reflecting lamp or lantern.

8. If worms be kept in confinement in May or June they may easily be observed in the act of copulation, and will lay numerous eggs.

D. **Dissection.** (See Appendix.)

1. Pin out and lay open, under 50% alcohol, a freshly-killed worm, and make out the general arrangement of—

    a. Body-wall, alimentary canal, dissepiments, cœlom.

    b. Dorsal vessel, circular vessels.

    c. Testes, seminal vesicles.

    d. Nephridia.

    e. Make a sketch showing these parts.

    f. Dissect away part of the alimentary canal behind the clitellum, and observe the *ventral nerve-chain* below it; also the *sub-intestinal vessel.*

    g. Dissect out the alimentary canal, make out all its different divisions; lay open the stomach-intestine from the side, and observe the typhlosole, p. 127.

    h. Examine with ¼ inch objective the chloragogue-cells mounted in water.

    i. Examine likewise the contents of the calciferous glands, run in dilute acid, and note the effect.

2. Take a fresh specimen and work out the circulatory system.

    a. Study the cœlomic fluid with the microscope (¼-inch obj.).

    b. Trace out the dorsal vessel to its anterior end.

    c. Examine the circular vessels as follows :—

        α. Of the stomach-intestine.

        β. Of the gizzard and crop.

        γ. The aortic arches.

        δ. Branches to the calciferous glands.

    d. Make out the sub-intestinal vessel and compare its relations behind the gizzard, in the crop-region, and in the œsophageal region.

    e. Cautiously chloroform a very large worm, pin out under normal fluid, cut open dorsally with sharp

scissors.   Observe the pulsations of the dorsal vessel
and of the aortic arches.

 f. With a lens observe—
  α. Dorso-lateral vessels.
  β. Ventro-lateral vessels.
  γ. Supra-neural trunks.
  δ. Sub-neural trunk seen through the nerve-chain.
  ε. The beautiful red capillary network extending over
   the body-wall, dissepiments, etc.

 3. In the same specimen work out the nephridia.
  a. Study *in situ* with hand-lens.
  b. Dissect out a number (half a dozen or more) of entire
   nephridia with fine forceps and scissors.   Mount in
   normal salt solution, and examine with ¼-inch obj.
   Observe the ciliary action and the various portions
   of the nephridium.
  c. Remove a number of nephridia from a preserved specimen, stain with borax-carmine or alum-carmine (under the direction of the demonstrator), and mount
   in balsam.   Study carefully with low and high
   powers.

 4. Lay open a preserved specimen, dissect away the alimentary
canal behind the pharynx, and expose the ventral nerve-chain.
Observe—
  a. The ganglia and their connecting commissures.
  b. The lateral nerves.
  c. Trace the chain forwards to the œsophageal collar;
   notice the lateral nerves at this point.
  d. Follow the collar around to the dorsal side of the
   pharynx, and observe the cerebral ganglia and their
   nerves.
  e. In a large preserved specimen make out the pharyngeal
   ganglia.
  f. Cut open a preserved worm on the dorsal aspect and
   soak in 10% nitric acid for several days (i.e., until
   thoroughly macerated).   The entire nervous system
   can then be easily removed, and after cleaning and
   dehydration, etc., may be mounted in balsam.   The
   specimen should not be stained.

 5. In a freshly killed and opened worm (preferably in June),

cut through the stomach-intestine, seize the fore part at the cut
with forceps and turn it forwards, gradually cutting away its
connection with fine scissors so as finally to expose the reproductive organs. The smaller parts must be sought out with a hand-
lens. Observe and sketch —

 a. Testes and seminal vesicle.

 b. Seminal receptacles.

 c. Sperm-ducts—follow them out.

 d. Ovaries. Remove with fine forceps, mount, and examine with the microscope.

 e. Oviducts.

 f. Examine with high powers spermatozoa from the seminal vesicles or seminal receptacles.

E. **Histology**. (See Appendix.)

 1. *Preliminary study.* With a sharp razor make thick sections (i.e., about $\frac{1}{16}$ inch) of a preserved specimen through the
following regions, viz., pharynx, œsophagus, crop or gizzard,
clitellum, and at two or three points of the region behind the
clitellum. Place the sections for an hour or more in strong
glycerine, mount in glycerine and study with low power (1-inch
obj.). Make out and sketch the general arrangement of parts,
and as many details as possible. Observe especially the setæ,
the muscles, and the alimentary canal. Preserve these slides for
comparison with the following :—

 2. Make thin cross-sections by the paraffin-method through
the same regions studied in (1). Observe with low power—

 a. Layers of the body-wall.

 b. Layers of the alimentary wall, typhlosole.

 c. Position of blood-vessels, nerve-chain, nephridia, setæ,
  and other organs.

 d. Interpret in the sections the appearances of the dissepiments, often very puzzling on account of their
  oblique position.

 e. Observe the stained corpuscles of the cœlomic blood—
  often very clean and beautiful.

 f. Study with high powers the structure of—

  α. Body-wall.

  β. Alimentary wall.

  γ. Nerve-ganglion. Trace nerves out into muscles.

  δ. Nephridia, reproductive organs, etc.

 g. Compare with the preceding a longitudinal section.

The microscopic study of the earthworm may be almost indefinitely prolonged, and is of great interest. The foregoing outline indicates, however, all that is necessary for the beginner. The study of the embryological development is impracticable for the beginner under ordinary circumstances, though at the proper season and with suitable guidance it may be made interesting and valuable.

CHAPTER XI.

OUTLINES OF CLASSIFICATION.

It is evident to the most superficial observer that there are certain likenesses among living things in accordance with which they may be arranged or classified into groups of greater or less extent. These likenesses are such that a study of any member of a group affords a certain amount of knowledge of all the other members. If we carefully study the structure and actions of a bird we learn a large number of facts which are true of all birds, however different they may be. In like manner the study of a fern, earthworm, oyster, moss, fish or oak, imparts not only a special knowledge of these particular organisms, but also a more general knowledge of all other plants and animals. This all-important fact renders it possible to acquire a knowledge of vital phenomena, in spite of the endless diversity of living things, through the study of a comparatively small number of forms selected as types, as will be attempted in the second part of this book. Were it otherwise, the student of biology would be lost in a maze of details; and indeed a science of living things would be impossible.

A closer examination shows that larger and smaller groups can be distinguished, each larger group consisting in general of a number of smaller ones, and forming in its turn a single member of a still larger group, and so on.

The earthworm, for instance, is a member of a small group termed a *genus* (*Lumbricus*) consisting of a number of different kinds or *species* of earthworms (*agricola, foetidus*, etc.). But the group of earthworms (*Lumbricus*) is a member of a larger group comprising all worms (*Vermes*), and the Vermes constitute in their turn a member of the still larger group *Metazoa*, comprising all animals whose bodies consist of many cells. The group Metazoa is in turn a member of the group of *animals*, which finally forms one subdivision of *living things*.

The fern, *Pteris aquilina*, is similarly a member of the genus known as *Pteris*, which includes many species (*aquilina*, *hastata*, *cretica*, etc.). *Pteris* is but one of many genera which are associated to form a larger group, *Vascular Cryptogams* (often called *Pteridophyta*), which is one member of the great group *Metaphyta*, including all many-celled plants; and this in turn is a member of the group of plants in general.

In this manner the great primary group of living things may be successively subdivided into secondary, tertiary, quaternary groups, etc.; and the study of any single organism, as a bird, therefore, teaches us not only a great deal of the smaller group, *Aves*, to which it immediately belongs, but also much of the larger group *Vertebrata*, which includes the *Aves*, and something of the still broader group, *Metazoa*, and of *Animalia* in general. Thus regarded, every living thing is full of meaning as a representative of other living things more or less nearly like itself.

Basis of Classification.   Living things may be classified by their resemblances in structure, in function, or in both. But experience has shown that no consistent classification can be founded on structure alone, or on function alone, and that both factors must be considered in any natural system. It has been pointed out at the close of the last chapter that the only fundamental difference between the fern and the earthworm lies in the mode of nutrition—in other words, it is a difference of function. This particular plant and animal are to a great extent typical of plants and animals generally. Broadly speaking, all plants agree with the fern in being able to construct proteid matter out of much simpler compounds, while all animals lack this power, and must be supplied with proteid matter as part of their food. In the last analysis this is the only point which will distinguish all plants from all animals. And hence the arrangement of all living things into the two primary groups of plants and animals is based upon differences of *function*, and not of *structure*.

The case is different, however, with the subdivisions of plants and animals; for, with few exceptions, it is found necessary to found these upon peculiarities of structure (using this word to include development) rather than of function. For example, birds and butterflies agree functionally in possessing the power of flight by means of wings. But the wings of the two animals are totally unlike in structure, although performing the same function.

For this reason the bird and the butterfly are not placed in the same group, since it can be shown that the likeness in function is less important than the difference in structure. The same principle holds true of nearly all kinds of plants and animals, and hence biologists are agreed that a natural classification must rest in the main upon structural characters; that is, it must be a *morphological* classification. There are, however, certain doubtful cases (e.g., *Bacteria*) in respect to which biologists are not yet agreed whether structure or function forms the most available basis of classification.

Natural and Artificial Systems. Morphological likeness or difference is therefore the usual basis of classification, but it is important to determine what particular likenesses or differences shall be taken as characteristic of the various groups. It is easy to show by a single example that morphological characters have very unequal values when regarded from this stand-point. It would be possible to classify all living things according to color, as white, yellow, green organisms, etc. Such a classification would, however, be artificial and destitute of scientific value because based upon a purely superficial and highly inconstant character. An interesting example of an artificial classification formerly employed is the system of Linnaeus, who classified flowering plants into *Monandria, Diandria, Triandria, Tetrandria,* etc., according to the number of stamens. This was sufficiently convenient for a first, rough arrangement, but was soon found to lead to the most incongruous association of plants agreeing in the number of stamens, but differing in almost all other characters. From such cases it is plain that plants and animals cannot be naturally classified by likeness or difference in a single character arbitrarily selected. The entire organism must be taken into account, and the natural classification differs from an artificial one in representing real relationship, and not merely a superficial likeness. Modern biology teaches that this relationship is of precisely the same kind as human relationship, i.e., that it is due to *community of descent* from ancestral plants or animals.

The origin of plants and animals by descent from other preexisting forms can no longer be questioned; and it is evident that the origin of any given plant or animal has been a definite process, to be determined like any other fact of natural history. It is plain, therefore, that there can be but one natural system of

classification, and in this system every plant and animal has its
fixed place, with definite relationships to every other plant and
animal. It is equally plain, however, that the natural system of
classification cannot be fully laid down until the ancestral history
of plants and animals has been completely ascertained. This
task has not yet been accomplished, and any attempt to formulate
the natural system must therefore be at present only approxi-
mate and tentative. It is natural that authorities should differ
more or less widely in these attempts, and hence a number of
systems are now in vogue, agreeing in their main outlines, but
differing considerably in details.

Correlation of Structure. The labor of determining the natural classification is much lightened by the fact that certain structures are often found as a matter of experience to be constantly
associated or correlated, so that the presence of one indicates the
presence of the others. In such cases a single character may be
taken as the basis of a classification which is natural, because
agreement in the one character has been proved empirically to
indicate agreement in many others. For example, it has been
proved that the differences or resemblances in the structure of
mammals are almost invariably correlated with corresponding
differences or resemblances in their teeth. Hence mammals, to
a great extent, can be naturally classified according to the structure and disposition of the teeth. And so in any group it is
usually possible to discover empirically some one or few characters on which, by reason of their constant association with other
characters, a natural classification can be based.

Homology and Analogy. A structural similarity due to common descent is known as *Homology.* Examples of mutually
homologous organs are the wing of a bird, the fore-leg of a dog,
the flipper of a seal, and the fore-limb of a mole. Although
these organs perform utterly different functions they exhibit a
fundamental likeness or homology in structure, due to the fact
that these different forms of fore-limb are descended from the
fore-limb of a common ancestor, their divergence in function
being a matter of secondary importance.

On the other hand, a similarity in function, whether accompanied or not by a similarity of structure, is known as *Analogy.*
The wing of a bird and the wing of a bat are not only homologous in structure, but also analogous in function. The wing of

a bird and that of a butterfly are likewise analogous in function, but they are not homologous in structure because not derived from a common ancestor.

Limits of Groups in the Natural System. Although living things can thus be arranged in natural groups which represent deep-seated similarities or relationships due to community of descent, it must not be supposed that these groups are absolutely distinct. On the contrary, the progress of biological knowledge has shown that the various groups of plants and animals merge into one another more or less completely; that there are connecting links between them, sometimes so numerous and so nicely graduated that it is impossible to draw any hard and fast line between groups which, broadly considered, appear very distinct.

Many of the earlier biologists maintained the various groups of plants and animals to possess well-defined boundaries, and denied the existence of real connecting links. Cuvier held that all animals without exception might be arranged in four absolutely separate great natural groups—the *Radiata, Mollusca, Articulata,* and *Vertebrata.* By some of his pupils this conception was extended from the four primary groups to their smallest subdivisions, so that even genera and species were considered immutable and distinct, in accordance with the view of Linnæus. At present, however, few if any biologists maintain that the groups, whether smaller or larger, are in reality separated from one another by any well-marked boundaries. Some of them appear to be so at first sight, but in many cases a careful examination of these groups shows them to be connected by intermediate forms, either living or extinct, and we may well hesitate to assume that this is not true of all groups. For example, birds and reptiles form two very distinct natural groups with no living connecting links between them. But the mesozoic rocks contain the fossil remains of animals which must be regarded as intermediate between birds and reptiles; and therefore the boundary-line between them cannot be regarded as impassable. There are strong reasons for believing that species, like the larger groups, have no absolutely fixed limits, but pass into related species by insensible gradations. These gradations are, in some cases, found among living forms, but in others the connection can only be demonstrated by the study of those which have become extinct. The extinction of the connecting forms, in many cases,

12

is what might be expected according to the view that species, like larger groups, have been derived by the modification of pre-existing ancestral forms.

The foregoing facts do not obviate, but rather emphasize, the necessity of an arrangement of plants and animals in a system of classification according to relationship. The groups, whether larger or smaller, have a real existence in nature as the expression of resemblances due to community of descent. But it must be clearly understood that they are not isolated and unrelated entities, but simply assemblages of individuals which group themselves naturally about certain centres, but shade off more or less gradually into other groups.

Outline of the System. As a matter of practical convenience, our primary classification of living things may differ according as the point of view is mainly physiological or morphological. For the sake of convenient reference in the present work, which will consider living things from both points of view, the outlines both of a physiological and a morphological grouping are here given.

GENERAL CLASSIFICATION ACCORDING TO STRUCTURE.

LIVING THINGS.	ONE-CELLED ORGANISMS.	Protophyta (plants). Protozoa (animals).
	MANY-CELLED ORGANISMS.	Metaphyta (plants). Metazoa (animals).

This grouping is extremely convenient, but there is no well-marked limit between one-celled and many-celled organisms, as will appear farther on. Moreover, as has been seen in the fern and earthworm, every many-celled organism during one stage of its existence (germ) is one-celled.

GENERAL CLASSIFICATION ACCORDING TO FUNCTION.

		CHARACTERISTICS.
LIVING THINGS (deriving energy from the sun).	GREEN PLANTS.	1. Able to absorb the kinetic energy of sunlight and to fix it as potential energy in the construction of complex matter, including proteids; therefore, 2. Not requiring proteid matter or other energized foods.
	COLORLESS PLANTS (Fungi, etc.).	1. Not able to absorb the kinetic energy of sunlight, requiring energized foods, but 2. Not requiring proteid matter.
	ANIMALS.	1. Not able to absorb the kinetic energy of sunlight, and 2. Requiring proteid matter as well as other energized foods.

This classification will be found useful in physiological discussions, but does not represent relationship in the sense of community of descent. For example, there are certain fungi which are very closely similar in structure to green plants, and have probably shared their ancestral origin. But if they ever had the power of utilizing the energy of sunlight in constructing complex food, they have lost it, and hence must be assigned to an entirely different physiological group.

In the detailed study of the relationships of plants and animals it is necessary to adopt a classification based as nearly as our knowledge will permit upon the resemblances due to community of descent. For the sake of convenient reference the classification here adopted and shown in the following table is nearly identical with that of two leading authorities in Botany and Zoölogy, viz., *Sachs* and *Claus*, to whose works the student is referred for further details.

SPECIAL CLASSIFICATION ACCORDING TO SACHS AND CLAUS.

LIVING THINGS.			
PLANTS. *(Sachs.)*		**ANIMALS.** *(Claus.)*	
Type.	*Class.*	*Type.*	*Class.*
I. **Thallophyta** (lowest plants)	Protophyta Zygosporeæ Oösporeæ Carposporeæ	I. **Protozoa** (lowest animals)	Rhizopoda Infusoria
II. **Muscineæ** (mosses, etc.)	Hepaticæ Musci	II. **Cœlenterata** (polyps, sponges, jelly-fish, etc.)	Spongia Anthozoa Polypomedusæ Ctenophora
III. **Vascular Cryptogams** or **Pteridophyta** (ferns, etc.)	Equisetineæ Filicineæ Dichotomeæ	III. **Echinodermata** (star-fish, sea-urch- ins, etc.)	Crinoidea Asteroidea Echinoidea Holothuroidea Enteropneusta
IV. **Phanerogamia** (flowering or seed- bearing plants)	Gymnosperms Angiosperms [Monocotyledons] [Dicotyledons]	IV. **Vermes** (worms)	Platyhelminthes Nemathelminthes Annelida Rotatoria
		V. **Arthropoda** (crabs, insects, spi- ders, etc.)	Crustacea Arachnida Onychophora Myriapoda Insecta
		VI. **Mollusca** (snails, oysters, etc.)	Lamellibranchiata Scaphopoda Gastropoda Pteropoda Cephalopoda
		VII. **Molluscoidea** (moss-animals, etc.)	Bryozoa Brachiopoda
		VIII. **Tunicata** (ascidians, etc.)	Tethyodea Thaliacea
		IX. **Vertebrata** (back-boned ani- mals)	Pisces Amphibia Reptilia Aves Mammalia

APPENDIX.

INSTRUMENTS AND UTENSILS.*

The student should be provided with the following articles:—

A compound microscope with two eye-pieces and low and high power objectives (i.e., about 1 in. and ⅓ in., or objectives A and D of Zeiss; still higher powers are desirable).

A simple dissecting microscope; a desirable form is an ordinary watchmaker's lens provided with a support. An ordinary pocket-lens; glass slides (3 × 1 in.), cover-glasses, watch-crystals, small gummed labels, needles with adjustable handles, camel's-hair brushes, blotting and filter paper, a good razor, pipettes (medicine-droppers), glass rods and tubes, glass or porcelain dishes for staining, etc., a set of small dissecting instruments (small scalpel, forceps, and straight-pointed scissors), a section-lifter, pieces of pith for section-cutting, thread, a shallow tin pan lined with wax, long insect-pins for pinning out dissected specimens, drawing materials, and a large note-book for sketches and other records.

Each table should be furnished with a set of small reagent-bottles, a Bunsen burner, wash-bottle, test-tubes, beakers, and a bell-glass for protection from dust. Thermometers, a balance, microtome, drying oven, and paraffin water-bath should also be accessible.

REAGENTS AND TECHNICAL METHODS.†

Alcohol.—Since biological laboratories belonging to incorporated institutions obtain alcohol duty free, it should if possible be *liberally*

---

* Most of the apparatus and reagents here mentioned may be obtained from any first-class dealer in physical and microscopical apparatus, e.g., from the Educational Supply Co., 6 Hamilton Place, Boston, or James W. Queen & Co., 924 Chestnut Street, Philadelphia. Chemical apparatus may be obtained from Eimer & Amend, 205–211 Third Avenue, N. Y.

† Every laboratory should be supplied with some of the standard books upon this subject, e.g., Strasburger's *Botanische Practicum*, Jena, 1884, and Whitman's *Methods of Research in Microscopical Anatomy and Embryology*, Boston, 1885.

supplied and freely used.   Alcohol of 100°, i.e., "absolute" alcohol, may be purchased in 1-pound bottles.   "Squibb's" absolute alcohol may be obtained of any druggist,* but ordinary alcohol of 90–95% answers nearly every purpose.   "Cologne spirits," i.e., alcohol of about 94°, may be obtained from the makers at 60c., or thereabouts, per gallon.   This may then be diluted to 80%, 70%, 50%, etc., as needed.   For this purpose an alcoholimeter is very convenient.

Acetic Acid.—One or two parts glacial acetic acid to 100 parts water.

Aniline Blue.—Dissolve the powder in distilled water.   It is best to dilute the solution largely when used.

Aniline Chloride.—A saturated aqueous solution slightly acidulated with hydrochloric acid.   Turns lignified cell-walls bright yellow (test for lignin).

Borax-carmine.—Add to a 4% aqueous solution of borax 2–3% carmine, and heat until the carmine dissolves.   Add an equal volume of 70% alcohol, and filter after standing 24 hours.   After staining (6–12 hours, or more for large objects; a few minutes for sections) place the object in acidulated alcohol (100 c.c. 35% alcohol, 3–4 drops hydrochloric acid) and leave until the color turns from dull to bright red (10–30 m.).   Afterwards remove to 70% alcohol.

Canada Balsam, Mounting in.—This invaluable substance may be obtained in the crude condition, dried by prolonged heating, and then dissolved in chloroform, benzole, or turpentine, for use.   The benzole solution is perhaps the best, and may be obtained from most of the dealers.   The principles of mounting in balsam are very simple.   It does not mix with water or alcohol, but mixes freely with clove-oil, chloroform, benzole, etc.   Objects are therefore generally treated, first with very strong alcohol, 95–100%, in order to remove the water; then with clove-oil, chloroform, or turpentine to remove the alcohol, and afterwards mounted in a drop of balsam.   This should usually be placed on the cover-glass, which is thereupon inverted over the object.   The balsam gradually sets firmly and the preparations may be permanently preserved.

Carmine.—Carmine may be obtained as a powder, which when rubbed up thoroughly with water in a mortar passes into a state of very fine subdivision.   This property makes it available for experiments with cilia, etc.

It is more often used in solution, as a staining agent.   (See **Borax-carmine.**)

---

* See also Whitman, l. c., p. 14.

Cellulose-test.—Saturate the object in iodine solution, wash in water, and place it in strong sulphuric acid prepared by carefully pouring 2 volumes of the concentrated acid into 1 volume of water.

Collodion and Clove-oil.—Used for fixing sections to the slide in order to prevent the displacement of delicate or isolated parts in balsam-mounting. Mix one part of ether-collodion and three parts of oil of cloves. In mounting, varnish a slide with the mixture by means of a camel's-hair brush, lay on the sections, and place the slide for a few minutes on the water-bath (i.e., until the clove-oil evaporates). Transfer the slide to a wide-mouthed bottle of turpentine (to dissolve the paraffin), remove it and drain off the turpentine, place a drop of Canada Balsam on the middle of a cover-glass, and invert it over the object.

Eosin.—Dissolve in water until a bright-red solution is obtained. It should be diluted when used.

Glycerine, dilute.—Two parts glycerine, one part distilled water.

Hæmatoxylin (Grenacher's).—Add 4 c.c. of saturated alcoholic solution of hæmatoxylin to 150 c.c. of strong aqueous solution of ammonia-alum; let the mixture stand a week or more in the light, filter, and add 25 c.c. of glycerine and 25 c.c. of methyl alcohol. The fluid improves greatly after standing some weeks or months.

Hæmatoxylin (Kleinenberg's).—To a saturated solution of calcium chloride in 70% alcohol add an excess of *pure* alum; filter after 24 hours and add 8 volumes of 70% alcohol, filtering again if necessary. Add a saturated alcoholic solution of hæmatoxylin until the liquid becomes purple-blue. The longer the liquid stands before using, the better. It should be diluted for use with the alum-calcium-chloride solution in 70% alcohol.

Iodine Solution.—Dissolve potassium iodide in a small quantity of water, add metallic iodine until the mixture assumes a dark-brown color, and then dilute to a dark-sherry color. The solution should be kept from the light.

Magenta (Aniline Red).—Dissolve in water.

Methyl Green.—Used in aqueous or alcoholic solution.

Normal Fluid (Normal Salt Solution).—Dissolve 6.75 grams of sodium chloride in 1 litre of distilled water.

Paraffin.—"Hard" and "soft" paraffins, i.e., those of high and low melting-points, should be mixed in such proportions that the melting-point lies between 50 and 55 C.

Perenyi's Fluid.—Ten-per-cent nitric acid 4 parts, 90% alcohol 3 parts, 1% aqueous solution of chromic acid 3 parts. Not to be used until the mixture assumes a violet hue. Leave objects in the fluid 30 minutes to an hour, then 24 hours in 70% alcohol, and finally place in 90% alcohol.

Schultze's Macerating Fluid.—Dissolve a gram of potassium chlorate in 50 c.c. of nitric acid. The tissue should be boiled in the mixture and afterwards thoroughly washed in water.

Schultze's Solution.—Dissolve zinc in pure hydrochloric acid, evaporate in the presence of metallic zinc, on a water-bath, to a syrupy consistency, add as much iodide of potassium as will dissolve, and then saturate with iodine. (When heated with this fluid cellulose turns blue.)

Section-cutting.—Many objects can be cut by hand with a razor (which must be very sharp). The object should be held in the left hand while the razor is pointed away from the body, and allowed to rest on the tips of the fingers with its edge turned towards the left. It is then drawn gently towards the body so as gradually to shave off the section. Small objects may be held between two pieces of watchmaker's pith previously soaked in water. In either case the razor should be kept wet.

Many objects, however, require more careful treatment by one of the following methods:—

A. *Paraffin Method.*—After hardening and staining, the object is soaked in strong alcohol (95% or more) until the water is thoroughly extracted (2–12 hours, changing the alcohol at least once), then in chloroform until the alcohol is extracted (2–12 hours), and then in melted paraffin (not warmer than 55° C.) on a water-bath for 15 to 30 minutes (too high a temperature or too long a bath causes excessive shrinkage). Some of the paraffin is then poured into a small paper-box, or into adjustable metal frames. The object is transferred to it and after the mass has begun to set it is placed in cold water until quite hard. It is then cemented (by paraffin) to a square piece of cork and placed in the section-cutter or microtome.

The sections may be cut singly with the oblique knife or by the ribbon-method,* the knife being kept dry in either case. In mounting they should be fixed by the collodion-method. (See **Collodion** and **Clove-oil.**)

B. *Celloidin Method.*—This is especially applicable to delicate vegetal tissues. After dehydrating the object thoroughly in alcohol, soak it 24 hours in a mixture of equal parts of alcohol and ether. Make a thick solution of celloidin in the same mixture and soak

---

* See Whitman, l. c., p. 71.

the object for some hours in it. It may then be imbedded as follows : Dip the smaller end of a tapering cork in the celloidin solution, allow it to dry for a moment (blowing on it if necessary), and then build upon it a mass of celloidin, allowing it to dry a moment after each addition. Transfer the object to the cork and cover it thoroughly with the celloidin. Then float the cork in 82–85% (0.842 sp. gr.) alcohol until the mass has a firm consistency (24 h.). It may then be cut in the microtome with the oblique knife, which must be kept dripping with 82–85% alcohol. Keep the sections in 82–85% alcohol until ready to mount them, then soak them for a minute in strong alcohol, transfer to a slide, pour on chloroform until the alcohol is removed, drain off the liquid, quickly add a drop of balsam, and cover. (See also Whitman, l. c., p. 113.)

OBJECTS FOR STUDY.

Most of the plants and animals used in the practical studies can be readily procured nearly everywhere, either in the field, in the market, or in hot-houses. We have usually been able to obtain them at reasonable prices from Mr. Wm. P. Seal, 205 South Eleventh Street, Philadelphia.

Anacharis, Chara, Nitella, Vallisneria, and *Spirogyra* (green pond-scum) occur in ponds or slowly running streams. *Tradescantia* is common in gardens and hot houses, and any of the cultivated species will answer. *Vaucheria* commonly forms dense green mats on the surface of flower-pots in hot-houses ; the larger species are aquatic, and are often associated with *Spirogyra*. *Amœba* may usually be found in the ooze at the bottom of bog-pools, particularly in *Sphagnum*-swamps. *Gregarina* may often be procured by stripping earthworms through the fingers, puncturing the swollen end, and examining the fluid which issues.

Pteris aquilina may be obtained in good condition between June and September. For demonstrating the general features and the coarser anatomy entire plants should be preserved in 50% alcohol, though dried specimens answer fairly well after being soaked some time in water. For histological purposes the terminal buds, parts of the leaf, and pieces of the rhizome from 1–3 inches long should be kept on hand preserved in 70% alcohol, to which they may be transferred after preliminary treatment for a few days with 50% alcohol changed every day or two. For fresh material in winter any of the hot-house forms of *Pteris* will answer, so far as the structure of the leaf and reproductive organs is concerned, but

these species are very different from *P. aquilina* in the shape of
the leaf and the habit of the stem. Alcoholic rhizomes furnish
excellent material for winter work, though fresh specimens are
preferable.

In the lack of *Pteris* almost any fern will answer for study,
though close agreement with the text must not be looked for.
Asplenium, Aspidium, Adiantum, the wild maiden-hair fern (*Adiantum pedatum*) or *Polypodium vulgare* may be used. Directions
for practical study of *Adiantum* will be found in the *Plant Dissection* of Arthur, Barnes and Coulter (Holt, N. Y., 1886), and of
Aspidium in the *Practical Botany* of Bower and Vines (Macmillan, 1885). Directions for procuring prothallia are given on p. 115.

Lumbricus.—Success in the practical study of the earthworm
depends very largely upon having *large specimens* properly preserved
by the simple method described below. If well preserved, earthworms are very satisfactory objects of study ; if not, they are worse
than useless.

Large earthworms may be obtained in great numbers between
April and November, by searching for them at night with a lantern, in localities where numerous castings show them to abound
(a rather heavy but rich soil will be found most productive).
They will then be found extended from their burrows, lying on the
surface of the ground, and may be seized with the fingers. Considerable dexterity is needed, and it is necessary to tread very softly
or the worms take alarm and instantly withdraw into their burrows.

For dissection both fresh and preserved specimens must be
used. Fresh specimens should be cautiously chloroformed under
a bell-glass (too long an exposure to the vapor or contact with the
liquid causes excessive contraction), then stretched out at full
length by pinning the fore and hind ends fast to a flat, wax-lined
vessel, opened by a *shallow* cut along the dorsal aspect, the flaps
pinned out, and the whole animal then covered with normal salt
solution or weak alcohol (40–50%).

For preservation (every detail of which should be attended to)
a number of living worms are placed in a broad vessel filled to a
depth of about an inch with water. A little alcohol is then cautiously dropped on the surface of the water at intervals of a few
minutes until the worms are stupefied and become perfectly motionless and relaxed (this may require three or four hours). They are
then transferred to a large shallow vessel containing just enough
50% alcohol to cover them, and are carefully straightened out and
arranged side by side. After an hour the weak alcohol is replaced

by stronger (70%), which should be changed once or twice at inter-vals of a few hours; they are finally placed in 90% alcohol, which should be *liberally used*. The trouble demanded by this method will be fully repaid by the results; and well-preserved specimens are in some respects better than fresh. They should be quite straight, fully extended, and plump, and they may be used either for dissection or for microscopic study.

For the purposes of section-cutting worms should be carefully washed and placed in a moist vessel containing plenty of wet filter-paper torn into shreds. The worms will devour the paper, which should be changed several times until the paper is voided perfectly clean.* The worms are then preserved in the ordinary way, and when properly hardened are cut into short pieces, stained with borax-carmine, imbedded in paraffin, and cut into sections with the microtome.

The embryological development is very difficult to study. In the latitude of Philadelphia egg-capsules may be found in great numbers in old manure-heaps in May and June. One end of the capsule should be sliced off with a very sharp scalpel and the con-tents drawn out, under water, with a large-mouthed pipette. The mass may then be mounted in water under a supported cover-glass, and studied with the microscope. The embryos may be preserved in Perenyi's Fluid, and either studied whole in the preserving fluid or hardened in alcohol and cut into series of sections.

-----

* This method of freeing the alimentary canal from earth and sand was suggested by Professor Benjamin Sharp.

INDEX

13